T0197937

Praise for *Technology Is Not Neutral*

Our globe-spanning economy, and our social interactions, depend on ever more pervasive digital technology, controlled by governments and multinational conglomerates. We're confronted by trade-offs between security, privacy and freedom. Stephanie Hare offers the overview that concerned citizens need to ensure that these potentially scary tools aren't misused. Her book deserves wide readership.

Professor Lord Martin Rees, Astronomer Royal and author of *On the Future: Prospects for Humanity*

A highly readable and enlightening introduction to the ethics of technology – with none of the usual finger-wagging! You'll never look at your cell phone the same way again.

Stuart Russell, Professor of Computer Science, University of California, Berkeley and author of *Human Compatible: AI and the Problem of Control*

Stephanie Hare makes an important and very timely contribution to our current debate over the power of Big Tech and the seemingly inexorable advance of artificial intelligence. Using telling examples from the past and the present she obliges the reader to consider the price humanity can pay for new technologies and how we can and must think ethically about their use.

Margaret MacMillan, Emeritus Professor of International History, University of Oxford

Stephanie Hare has addressed one of the biggest questions confronting us all – how we can create and use tech to maximize benefits and minimize harm – with great clarity, wisdom, and confidence. Drawing on the insights of numerous academic fields as well as concrete, real-world examples, this is an extremely useful guide to thinking about what we should ask of technology.

Adam Segal, Director, Digital and Cyberspace Policy Program, Council on Foreign Relations

Technology Is Not Neutral

PERSPECTIVES

SERIES EDITOR: DIANE COYLE

Technology Is Not Neutral

A Short Guide to Technology Ethics

Stephanie Hare

LONDON PUBLISHING PARTNERSHIP

Published by London Publishing Partnership
www.londonpublishingpartnership.co.uk

Published in association with
Enlightenment Economics
www.enlightenmenteconomics.com

ISBN: 978-1-907994-97-5 (hbk)
ISBN: 978-1-907994-98-2 (iPDF)
ISBN: 978-1-907994-99-9 (epub)

A catalogue record for this book is
available from the British Library

This book has been composed in Candara

Copy-edited and typeset by
T&T Productions Ltd, London
www.tandtproductions.com

Cover image by Noma Bar/Dutch Uncle

Printed and bound by TJ Books Ltd, Padstow

To my parents

Contents

Introduction

At first it was not clear that they were trying to stop the certification of the election and kidnap and kill lawmakers, including the vice-president. We only learned that later. They looked like peaceful protesters. Some of them were, but others were carrying weapons and planting explosives. They made their way from a 'Save America' rally near the White House and stormed the Capitol Building, prowling the corridors of the legislative branch of the most powerful country in the world – *their own country.* As they reached the chamber, lawmakers and police officers barricaded the doors. The police officers drew their firearms.

Some lawmakers sheltered in the chamber. Some had escaped to a secure location. Others hid behind locked doors in their offices, cowering with their colleagues in cupboards and under tables and desks. They held their breath, trying to stay silent as the mob banged on the doors and yelled at them to come out.

This went on for hours. Images and videos were recorded by journalists, witnesses, lawmakers and the rioters themselves. They were shared in real time. They went viral worldwide.

The man who inspired the attack, and who might have put a stop to it, stayed silent while the damage was done. Only later did he post a video on his Twitter account, which he had used for years to communicate directly to his 87 million followers. He urged his supporters to go home. He said he wanted peace.

Yet even as a defiant Congress worked through the night to certify the election, and even as law enforcement and security services began to piece together what had happened, President Donald J. Trump kept tweeting. He spread baseless accusations of election fraud. He denied that he had lost to President-Elect Joe Biden, whose inauguration was only weeks away. He attacked his own vice-president, who had been forced to flee and hide in a basement from a mob chanting 'Hang Mike Pence!'[1]

For two days Trump's tweets were shared across social media and mainstream media. Further violence loomed. It was unclear if anything could stop it. But then something – or rather some-one – did. On 8 January 2021 Jack Dorsey, then CEO of Twitter, cut Trump's microphone.[2]

Dorsey's decision to permanently suspend Trump's Twitter account sparked a chain reaction among other US technology leaders.[3] Facebook, Instagram (which is owned by Facebook), Snapchat and Twitch also took Trump off their platforms.[*] Amazon refused to host Parler – one of the apps used by Trump's supporters to organize the attack – on its web services, and Apple and Google removed the app from their app stores.[4] YouTube (which is owned by Google) deleted some, though by no means all, of Trump's videos.

This was a live demonstration of technology ethics in action, and so was what followed. The actions taken by those technol-ogy leaders ignited a debate in the United States and elsewhere. First, by what right had they silenced the president? After all, technology CEOs have no democratic legitimacy – they are only accountable to their shareholders. By contrast, the president is elected directly by the American people. Second, were the technology leaders preventing free speech – something that is protected by the First Amendment of the Constitution? Or did they have a right – maybe even a duty – to uphold their company policies that ban the glorification of violence and the risk of fur-ther incitement to violence?

---

[*] On 28 October 2021 Facebook changed its name to Meta, making its namesake service, Facebook, a subsidiary along with Instagram and WhatsApp.

Dorsey himself was unsure. In a long thread on Twitter he described his decision to ban Trump as 'right' but also 'dangerous'.[5]

jack ✦ ⦾
@jack

I do not celebrate or feel pride in our having to ban @realDonaldTrump from Twitter, or how we got here. After a clear warning we'd take this action, we made a decision with the best information we had based on threats to physical safety both on and off Twitter. Was this correct?

12:16 AM · Jan 14, 2021 · Twitter for iPhone

16K Retweets **12.2K** Quote Tweets **111.4K** Likes

Figure 1. First tweet in a long thread by Jack Dorsey, then Twitter's CEO (14 January 2021).

Whether Dorsey's actions, and those of the other technology companies that de-platformed Trump, were correct or dangerous – or should be judged by any other criteria we might care to choose – is a question of values, or *ethics*, and as such we are unlikely to reach consensus over these issues anytime soon. By contrast, enforcement of the *law* is well underway. Hundreds of people have already been arrested over what the director of the Federal Bureau of Investigation (FBI) has called an act of domestic terrorism and the US Department of Justice has described as 'likely the biggest criminal investigation in US history'.[6]

What happened on 6 January 2021 is one for the history books – but not *just* for the history books. It was a turning point for technology, too.

The attack was organized online, raising questions about whether the technology companies had a responsibility to flag it to the authorities. Many of the attackers had been radicalized online, across several platforms, long before the insurrection. Those who used the platform Dlive to livestream the attack *made money while doing so* because Dlive allows viewers to pay users who are broadcasting content.[7]

Technology has also played a crucial role in investigating the attack. The people who posted photos, videos and other information relating to the attacks created a rich source of data that law enforcement and the House Select Committee are using in their investigations. The data set includes, but is not limited to, mobile phone, photo and video analysis as well as facial recognition technology.

Members of the public have also conducted their own investigations using social media and facial recognition technology to crowdsource their 'sedition hunt'.[8] On the one hand, this is no different from law enforcement's usual request to the public to report any tips in ongoing investigations. Yet it also creates new risks of online vigilantism – what Eliot Higgins calls 'digilantism' – which could cause harm by misidentifying people.[9]

Congress – whose members failed to agree to create an independent commission to investigate the violence that threatened their lives and the certification of the election – is now more united than ever in the belief that technology companies are too powerful and need to be reined in.[10] By June 2021 it had introduced five bills to break up Big Tech, each one co-sponsored by Republicans and Democrats.[11]

President Biden seems sympathetic to their views and has appointed Lina Khan, a prominent critic of Big Tech, to lead the US Federal Trade Commission (the FTC – the United States's anti-competition regulator), and Jonathan Kanter, an antitrust lawyer who has spent his career taking on US technology giants, as the Justice Department's assistant attorney general for the antitrust division.[12] Even before the attack, Biden had said that he wanted to scrap Section 230 of the US Communications and Decency Act (1996), which protects freedom of expression on the internet.[13] If he chooses to pursue this, he will find support in Congress, where lawmakers on both sides of the aisle have complained.

The US government is not alone in looking to curb the power of technology companies. Even before the Capitol attack, the European Union had launched two landmark pieces of legislation: the Digital Markets Act (DMA), which would allow the European Union to break up technology companies, or at least make them sell off their European operations if they are judged to be too dominant; and the Digital Services Act (DSA), which would require online platforms to take down illegal content or counterfeit goods or be subject to substantial fines.[14] In the United Kingdom, the Competition and Markets Authority is setting up a Digital Markets Unit to police technology companies' dominance. In China, President Xi Jinping has been strengthening the regulation of his country's $4 trillion technology industry, including the introduction of new anti-monopoly rules, protections for gig workers, data protection laws, rules governing the role of algorithms in content distribution, and restrictions on the number of hours children under the age of 18 can spend gaming.[15]

These regulatory actions could damage or even devastate the operating models of many technology companies, not just the giants, so those companies will do everything they can to block them or at least weaken them as much as possible.[16] A battle looms. That is because technology is not just about tools and toys, products and services, data and code. It is about power. How we approach that power is shaped by our values – our ethics – and that is the focus of this book.

HUMANS, NOT COGS IN THE MACHINE

'I could probably write a very good program for choosing people to be killed for some reason, selecting people from a population by a particular criterion,' Karen Spärck Jones, a computer scientist and professor at Cambridge University, told the British Computing Society in 2007. 'But you might argue that a true professional would say, "I don't think I should be writing programs about this at all."'[17]

Spärck Jones was ahead of her time in more ways than one. She was one of those thinkers who, while celebrated in her field, is barely known outside it, and yet we rely on her work in statistics and linguistics every time we use a search engine.[18] She urged us to think about the ethics of what we create: 'You don't need a fundamental philosophical discussion every time you put finger to keyboard,' she said, 'but as computing is spreading so far into people's lives, you need to think about these things.'

Who should do this thinking?

'Computing is too important to be left to men,' Spärck Jones said, reflecting on her years of work in opening up that male-dominated field to all talent. I agree, and I would go even further: technology is too important to be left to technologists. We need everyone to hold technology to account.

Unfortunately not everyone may feel inclined to accept this invitation. Even the word 'technology' turns some people off. They hear it and their minds drift off elsewhere in order to escape talk of gadgets and code, hardware and software, and sci-fi references that only make sense to those who have read the books and seen the films. This turn-off can baffle technology enthusiasts, for whom the word 'technology' is a turn-on, triggering reactions such as the joy and satisfaction that come from solving problems, making life easier, finding new ways to have fun, shared cultural references, and plotting paths to wealth and power.

This is a dangerous divide. After all, technology is part of what makes us human. Only some of us *create* technology but we all *use* it, and we all have it used on us, sometimes without our knowledge and consent. Technology is at the interface between citizens and governments, between consumers and companies, and between humans and the planet. It is an essential element in understanding our history, our present and our future.

Even when *we* are not interested in technology, technology is interested in *us*. Many of the most valuable and influential companies in the world are technology companies. A growing number of governments have a digital services arm, the job of which is to create ways for us to access health, tax, passport and other services online, minimizing the need to set foot in any physical premises. Companies and governments are also working together to create digital identities for us, make our cities 'smart', and transform our civilian and defence infrastructure into a hybrid of the physical and the digital.

To ignore technology is a decision – one that turns us into a cog in someone else's machine. This decision places us at the mercy of the 'true professionals', hoping they will not harm us. Why would any of us accept such a passive role in our own lives when we can hold those who work with technology to account so that they work *for* and *with* us, rather than *against* us?

WE BEGIN AS DATA

To claim our power requires a mental shift – one that changes how we see ourselves.

For example, through the lens of physics, we are elemental. As Sir Martin Rees, the United Kingdom's Astronomer Royal, explains: 'We ourselves and everything in the everyday world are made from fewer than 100 different kinds of atoms – lots of hydrogen, oxygen, and carbon; small but crucial admixtures of iron, phosphorus, and other elements.'[19]

Through the lens of biology and chemistry, we are what happens when our mother's egg and our father's sperm combine to form a cell, with twenty-three chromosomes usually coming from each parent and fusing into pairs for a total of forty-six. Each chromosome is made of a long strand of DNA, which is divided into segments called genes. Each gene contains the information our bodies will need to grow and maintain themselves throughout our lives. Our 'code' is unique to us. Even identical twins do not have completely identical DNA.

Through the lens of the social sciences, the liberal arts and the humanities, we are not simply elements involved in biological and chemical processes who are born, reproduce and die. We are also social creatures who exist in a network of relationships and in history, not in isolation. We are the product of our environment, our experiences and our choices. Some of this is out of our control, some of it is down to us. We, in turn, both individually and collectively, shape the environment, experiences and choices of others.

Through the lens of technology, we are creators and users of tools, methods, processes and practices. We can have these things used on us. We are data that can be turned into computer code and then analysed to find things out about us – in the past and present, and to predict our future. Of course, not everything about us can be expressed as code or, indeed, be known. Some parts remain a mystery to those closest to us, and even to ourselves. Yet we can also be known in stunning detail by governments, companies, researchers and many others. By knowing us they can also know about our families, friends, colleagues, acquaintances and neighbours. By collecting, analysing and storing our data, they can sell to us, influence us, spy on us, manipulate and control us, and through their repeated failures to keep our data secure, they can expose us to risk from criminals and hostile nation states.

he problems are clear. What is less clear is how to solve them. That is one of the reasons I wrote this book, but it is not the only one.

WHY I WROTE THIS BOOK

I could have used a book like this when I was at high school and when I was an undergraduate student in the 1990s, preparing to enter a world in which technology was going to shape my life and career in ways I could not fathom.

I really needed it when I began my first job in technology in 2000 – and in every technology role I have had since for that matter.

It would have helped greatly when I began working as a political risk analyst in 2010. Surely I would have grasped more quickly that technology is key to understanding human behaviour and thus the power dynamics that shape politics and markets, culture and climate, and so much more.

It would have been indispensable when I began analysing technology developments for the media, some of them fast moving, others slow and stealthy, all complex and interconnected.

Most of all, I wish I had had something like this when faced with ethical dilemmas involving what technology to use, create and invest in.

Necessity being the mother of invention, I began researching and discovered a wealth of work by academics, researchers and journalists. I also found people from all walks of life who are not only *thinking* and *talking* about technology ethics but *doing* it. Specifically, a new role – that of technology ethicist – is emerging in our economy, but its contours are still being shaped. Is it a technologist who works in ethics? An ethicist who works in technology? Can anyone call themselves a technology ethicist or is it an anointed position?

Rather than focus on what technology ethicists *are,* let us consider what they *do.* They might be trained in law, data science or philosophy, or they might be artists or designers. They might be employed by universities (and not just in the philosophy and computer science departments) or they might work in think tanks, in non-governmental organizations (NGOs), in companies or in any part of government. They might do open-source intelligence investigations into crime, terrorism and human rights abuses. They might be lawyers who take on cases relating to privacy, civil liberties, data protection, human rights and competition. Perhaps they infuse new meaning into existing roles – professor, researcher, data protection officer – or they may reflect new and more specific responsibilities, such as responsible AI lead, algorithmic reporter, or AI ethicist. They might sit in a team that is explicitly dedicated to ethics, such as Twitter's Machine

Learning Ethics, Transparency and Accountability Team. Or they might work in teams such as the 'responsible technology' team, the 'trustworthy AI' team or in teams that deal with legal affairs, risk, compliance or cybersecurity. They could have junior, middle management or senior responsibilities. They might sit on internal ethics panels or boards. Finally, they may be on the board as a Chief Privacy Officer, Chief Ethics Officer or Chief Responsible AI Officer.

I wanted to explore this emerging role of 'technology ethicist' to see what difference, if any, it is making to improve how we create and use technology. Fortunately, many other people who were also keen to better understand this question were generous with their time and thoughts.

To learn what can be done with our face and other body data in the United Kingdom, I have spoken with police officers from the London Metropolitan Police, the UK Biometrics Commissioner, the head of the Science and Technology Committee in the House of Commons, and a mother who campaigns against the widespread use of children's biometrics in UK schools.

I participated in separate workshops on technology ethics with one of the world's largest weapons makers and with one of the world's largest toy makers.

I talked with the former Chief Ethics Officer of Airbnb to learn what he means by 'intentional integrity', how organizations can foster it, and what risks they face if they fail to do so.

I spoke to the team of twelve-year-old girls who won the CyberFirst talent competition hosted by GCHQ (the United Kingdom's intelligence, security and cyber agency), to the talent spotter for the UK National Cyber Security Centre, and to three historians whose research has highlighted the pioneering role of women in cybersecurity.

I talked to historians of medicine, anthropologists, public health officials in the London borough of Hackney, and doctors about the problem of the pandemic and what role, if any, digital health tools could play in solving it.

Figure 2. Examples of technology ethics job adverts in the United States, Ireland and the United Kingdom, 2019–21.

I interviewed leaders of biometric technology companies, an immunologist at Imperial College, and a behavioural psychologist at University College London to explore the idea of Covid vaccine passports for domestic use.

I talked with a digital forensic scientist who helps solve crimes by examining the data on our devices, from our smartphones to our 'smart' washing machines.

Finally, I published opinion pieces in newspapers and magazines, gave public lectures and chaired and participated in panels. From this I received feedback from members of the public and from specialist audiences of civil servants, cybersecurity professionals, private wealth investors, urban planners, management consultants and telecommunications companies.

Out of all this research and discussion, a fundamental problem emerged: *technology ethics concerns all of us but we are not sure how to do it or how to know if what we are doing is working.*

A SHORT GUIDE TO TECHNOLOGY ETHICS

This book aims to address this by exploring what technology ethics is, how we can 'do' it, and how we can know if what we are doing is working. At its core is the question of how we can create and use technology and tools in a way that maximizes benefits and minimizes harms.

In the first half of the book we will explore a debate that sits at the heart of technology ethics: is technology neutral? From there, we will consider a question that arises whenever we create or use a tool or technology: where do we draw the line?

In the second half we will examine two problems that are global, complex, ongoing and increasingly impossible to ignore: the use of facial recognition technology and digital health tools for the Covid-19 pandemic. We will look at what problems they purport to solve and whether they even work; what they mean for our identity, privacy and civil liberties; how they change our experience of where and how we live; whether they are a good thing or a bad thing; and what we should do about them now that we have invented them.

We will conclude by looking ahead to how we might apply technology ethics to other challenges, risks and opportunities – of which there is no shortage.

Chapter 1

Is technology neutral?

> When you invent the ship, you also invent the shipwreck; when you invent the plane, you also invent the plane crash; and when you invent electricity, you invent electrocution... Every technology carries its own negativity, which is invented at the same time as technical progress.[1]
>
> — Paul Virilio, French cultural theorist and philosopher

Is technology neutral? This is one of the most contentious questions in technology today. It is more than an academic debate and more than a matter of personal opinion. How we answer it can help us to determine the responsibilities of anyone involved in creating or using technology. It can also help us solve the main question of this book: how can we create and use technology to maximize benefits and minimize harms?

Technology shapes our lives and those of other people with whom we come into contact, both directly and indirectly, whether we are aware of it or not. It can act as a liberating force but it also entrenches asymmetries of power across gender, race, class, generations and geography. It is at the core of some of the most valuable companies not only of today but in history – companies whose power and influence challenge the authority of governments around the world. It can help people hold their governments to account but it also helps those governments to surveil more people, in more detail, than ever before.

It is a matter of superpower rivalry between the United States, the European Union and China, often forcing other countries to take sides not just on technology suppliers but also on values. It challenges our ideas about what it means to be human. It forces us to reconsider what it means to have autonomy and to enjoy privacy, civil liberties and human rights.

Technology ethics concerns all of us – we all use technology and we all experience technology being used on us by others – but it *especially* concerns anyone involved in making technology, as they bear at least a degree of responsibility for their creations. That is why Joseph Weizenbaum, a computer scientist at MIT and an early pioneer of artificial intelligence (AI), issued this warning in 1987:

> It [is] possible not to know and not to ask if one is doing sensible work or contributing to the greater efficiency of murderous devices.
>
> One can't escape this state without asking, again and again, 'What do I actually do? What is the final application and use of the products of my work?' and ultimately, *'Am I content or ashamed to have contributed to this use?'*[2] [Emphasis added]

Nor can we hide behind good intentions, as Apple CEO Tim Cook explained in his commencement speech at Stanford in 2019:

> Too many seem to think that good intentions excuse away harmful outcomes. But whether you like it or not, *what you build and what you create define who you are.*
>
> It feels a bit crazy that anyone should have to say this, but if you built a chaos factory, you can't dodge responsibility for the chaos.[3] [Emphasis added]

He speaks from experience. Under Cook's leadership, Apple has refused to unlock iPhones for the FBI in criminal investigations; navigated a tricky path over privacy and data rights with China (where it assembles most of its products and makes around 20 per cent of its revenues); and unveiled a system that, while intended to scan users' iPhones for child sexual abuse material, could potentially be expanded to allow governments to scan private content on people's phones.[4]

In this chapter we will consider a debate between some people who think that technology is neutral and others who think it is not. Then we will examine various tools and technologies to evaluate how they compare on the question of neutrality. Finally, we will consider different forms of 'intelligence' and how this relates to decision-making and, ultimately, responsibility.

THE DEBATE

Imagine two teams composed of some of the most interesting thinkers in technology living today and ask them, 'Is technology neutral?' Here is what they might have to say.

Team 'technology is neutral'

Technology is neither good nor bad; it depends on how we use it. That is what Professor Daniela Rus, director of the Computer Science and Artificial Intelligence Lab (MIT's largest research lab), thinks. Here is how she explained her thinking at the World Economic Forum in 2019:

> I'm a roboticist. Now when I tell people what I do, I get one of two types of reactions. Some people get anxious. They make jokes about Skynet.* And they ask me, 'When will the

---

* For readers who have not seen the *Terminator* films, Skynet is a US-funded artificial intelligence warfare system that removes human decisions from strategic defence, becomes self-aware and starts fighting back against humanity. The humans send one man, John Connor, back in time to try to stop Skynet from being created. In response, Skynet sends the Terminator back in time to kill John Connor.

robots take over my job?' And then other people get very excited and ask me, 'When will my car be self-driving?'

Well, I belong to the second group. But I believe it's very important to understand the concerns of the first group and provide ideas and suggestions for how to see things differently. And this starts with understanding that AI and robotics and machine learning are tools. *They are just tools, by the people, for the people. They are incredibly powerful tools. But like any other tools, they're not good or bad. They are what we choose to do with them.*

And I believe we can choose to do incredible things. With AI, machine learning and robots we will ensure that there are no road fatalities. We will engineer better medicines. And we will better monitor and treat disease. We will ensure that we can transport people and goods faster and better. We will ensure that the world will be better connected, no matter what language is spoken. We will enable people to focus on strategic, creative thinking, with machines doing the low-level, routine tasks.[5] [Emphasis added]

Paul Daugherty, Chief Technology and Innovation Officer at Accenture, agrees. In April 2019 – while accepting on behalf of Accenture the award for the Corporate Honoree for Ethical Leadership from the Fellowships at Auschwitz for the Study of Professional Ethics – he said:

So I'd ask you to think about a simple question. Is technology, and Artificial Intelligence, good or bad?

The answer is NEITHER. *Technology is neutral, AI is neutral. The way 'we', as humans, apply and use the technology is what defines if the impact is good or bad.*

Why is this important? Artificial Intelligence is in our pockets, on our kitchen counters, in our vehicles. It's monitoring our bodies and movements; anticipating our interests, desires, and behavior; shaping our children's activity and experiences. And it's used not just by business but by our doctors, police, judges, educators.

Given the increasingly pervasive, and invasive, impact of technology on the way we work and live, ETHICS is no longer a peripheral issue in business, nor something you think about after the fact. The choices we make are critical. Ethics must be core to a company's strategy, culture, operations, and technology.[6] [Emphasis added]

Garry Kasparov, the chess grandmaster and author of *Deep Thinking: Where Machine Intelligence Ends and Human Creativity Begins*, also agrees.[7] In 2019 he posted on Twitter: 'Tech is agnostic, it amplifies us. "Ethical AI" is like "ethical electricity".'[8] Here 'agnostic' means that technology works across any platform, protocol or device without requiring any adaptation[9], the way that type O negative blood works for people of any blood type who need a transfusion. As such, Kasparov is focusing on the *physics* of electricity, which is neutral, rather than the *context* of electricity, which includes factors such as whether it is renewably sourced or produced by a company that has an ethical track record or is part of an ethical supply chain.

Amazon Vice-President and Chief Technology Officer Werner Vogels also argues that technology is neutral. He thinks that 'society' bears responsibility for the use of facial recognition technology, which Amazon creates and sells to many police forces around the world through a product called Rekognition:

That's not my decision to make. This technology is being used for good in many places.

It's in society's direction to actually decide which technology is applicable under which conditions...

Machine learning and artificial intelligence (AI) are like steel mills. Sometimes steel is used to make incubators for babies but sometimes steel is used to make guns.[10]

Demis Hassibis, the co-founder of DeepMind, a company that builds AI systems and was acquired by Google, seemed to agree in an interview with the BBC in 2019:

> I think AI is no different to any other powerful technology that was invented in the past. I think *the technology itself is neutral. The question is how the society decides to deploy it.*[11] [Emphasis added]

But he also said that scientists have a responsibility to understand the ethical implications of AI, which would suggest that technology is *not* neutral:

> *I do think scientists have a responsibility if they're working on cutting edge research that can have big impacts in the world, to think through the consequences of that ahead of time.* And what we try and do is to try and understand solutions, not just look at the performance or the outcome.
>
> You actually want to understand the process and that allows you to project forward or imagine forwards what the consequences of what you do might be. I mean, I've recently been arguing that we can't stop or maybe we can't even slow down the rate of technological progress in AI research.
>
> But the real issue is making sure we're ready for it when it comes. *We have to understand its implications, limitations, its risks, or the, you know, the ethical issues.* [Emphasis added]

Adam Mosseri, the head of Instagram, also appears to be in two minds about whether technology is neutral. For instance, five days after the 6 January 2021 attack in the United States he tweeted:

> *We're not neutral. No platform is neutral, we all have values and those values influence the decisions we make.* We try and

be apolitical, but that's increasingly difficult, particularly in the US where people are more and more polarized.[12] [Emphasis added]

Yet in August 2021 he tweeted:

Technology isn't good or bad, it just is. Social media has amplified good things, like #MeToo and #BlackLivesMatter, and bad things, like misinformation.[13] [Emphasis added]

Finally, Marc Andreesen, the billionaire Silicon Valley venture capitalist, said in September 2021 that 'technology is just a tool. It is human beings who decide how to use everything.'[14]

Yet not all humans get to decide how to use everything. Most of us will only ever be on the receiving end of other humans' use of AI to shape our lives: from whether we get called to interview to whether our mortgage application is accepted to whether we will be released from jail – and much more. We will explore this later in the chapter, but first we need to hear the other side of the 'Is technology neutral?' debate.

Team 'technology is not neutral'

Sir Tim Berners-Lee, who created the World Wide Web in 1989, argues that we must build ethical rules into the very design of the technology we are building:

As we're designing the system, we're designing society. Ethical rules that we choose to put into that design [impact society]... Nothing is self-evident. Everything has to be put out there as something that we think will be a good idea as a component of our society.[15]

That is one reason why he co-founded Inrupt in 2017, with the aim of changing the power dynamics of data ownership by reinventing the web.[16] As Bruce Schneier, the head of Inrupt's security architecture, explained:

> The problem is that your data is not under your control. It is on computers owned by lots of other people. And you have no control over it and you do not have access in ways that are useful to you.[17]

Both are highlighting what happens when technology crosses over from the theoretical to the real world. Where humans are concerned, neutrality is not only a question of values; it is a question of power.

That is why Professor Kate Crawford asked 'What is neutral?' during her lecture on the politics of AI at the Royal Society in 2018. 'The way the world is now? Do we think the world looks neutral now?' she said.[18] She suggests a way of testing whether technology is neutral: by considering its impact. 'Good or bad impact' is not, as Daugherty argues, defined simply by the way 'we', as humans, apply and use technology. For some technologies and tools, 'bad' impacts are a design *feature*, not a bug. They result from failures *embedded throughout the creation process,* not just during application and use.

For example, in *Invisible Women: Exposing Data Bias in a World Designed for Men,* Caroline Criado Perez shows how technological innovation often ignores women *at every stage,* from the mainly male composition of the teams that fund and create technologies to the absence (sometimes deliberate, often unthinking) of women's data in data science, product design and launch, and the legislation and regulation around it.[19] Sometimes the negative impacts of this failure to include women at every stage are simply annoying, such as products marketed as 'gender-neutral' that ignore the difference between average male and female hand sizes (e.g. smartphones) and voice and speech recognition technologies that function worse – or not at all – on women's voices, which reduces the effectiveness of voice assistants. Why, Criado Perez asks, should women pay the same as men for products and services that do not work as well for them – because they were designed for a 'default male'?

Design bias is sometimes worse than annoying: it is life-threatening. For example, speech recognition software used by attending emergency physicians has higher error rates for women than for men, endangering the lives of everyone in their care (men included). Voice recognition software in cars is sold as a technology that can make driving safer yet it does not work as well for women's voices. Both are a problem with the technology, not with women's voices. Alas, this point did not seem to dawn on Tom Schalk, vice president of ATX, a car voice technology system supplier, when he said that women 'need lengthy training' to use his company's tool.[20,21]

Such outcomes are worse than 'not neutral'. Women make up more than half the human population and design should reflect that.[22] But these non-neutral outcomes extend beyond women, too. As Sheila Jasanoff, professor of science and technology at Harvard University, explains, the difference in impact is not limited to how we *individually* experience technological innovation; it can also change our relationships *with one another*, and even *with our environment*.[23] 'The same technologies can be found from Kansas to Kabul,' she writes, 'but people experience them differently depending on where they live, how much they earn, how well they are educated, and what they do for a living.'

However, it is not enough to look at the technologies themselves; we must also consider the people who are creating them and infusing them with their values. 'Technology is not neutral. It's shaped by the people that build the technologies, shaped by their choices and their values,' warns Erin Young, a research fellow at the Alan Turing Institute. For example, 'there is mounting evidence that suggests the under-representation of women in AI roles within tech companies results in feedback loops. AI systems are not objective. So when bias goes in, bias comes out.'[24]

In her book *The Real World of Technology,* *the* experimental physicist Professor Ursula M. Franklin argues that we can do more than study the impact of a tool or technology to ascertain whether it is neutral: we can use the impact as the *starting point* from which to work backwards to understand why a tool or technology has been created in the first place.[25] Consider who is at the receiving end of a tool or technology, she suggests. For example, workplace surveillance technologies are used by managers, executives, the C-Suite and the board to manage and control workers, *not* vice versa. Similarly, surveillance technologies such as facial recognition are used by the police, governments and businesses to monitor and control the population, *not* the other way around.

The thinkers above believe that tools and technologies are instruments and expressions of power. To build them responsibly, we must consider the intent behind the creation of tools and technologies, the entire cycle from idea to execution, and the context in which we introduce them so that we can foresee the consequences – intended and unintended – and address them.

BETWEEN THE BONE AND THE BOMB

The argument that 'tools and technologies are neither good nor bad, it depends on how we use them' skips over an important difference between tools that are *found* and tools that humans have *created.* Let us consider two examples that illustrate the ends of a spectrum: a bone (a found tool) and the atomic bomb (a created tool).

The bone

In the film *2001: A Space Odyssey,* one of our ancestors, a hominin, sits in front of an animal skeleton. He stares at it for a while, then picks up one of the bones and has an insight: he can use the bone as a tool. As proof of concept he smashes the animal skeleton to bits. In the film's next scene he has shared the insight with the other hominins in the troop, and together they use their new tool as a weapon to defend themselves and their waterhole from an invading troop, going so far as to bludgeon one of the intruders to death.

Note that the hominins did not *create* the bone. Nor did they *modify* it in any way. They simply *found* it. They have no responsibility for its creation. The bone is a neutral tool: in this case, the hominins used the bone to defend themselves and to kill, but they could just have easily used it for other purposes.

The bomb

By contrast, the atomic bomb is not a found object. Humans had to create it, and that was very hard to do. It has only one purpose – mass destruction – whether it is ever actually exploded or if it is maintained as a threat of explosion (offensive) or kept as a deterrent against other people exploding *their* atomic bombs (defensive, leading to mutually assured destruction). The atomic bomb is therefore not neutral.

Where does responsibility for its creation and use lie? No one was trying to build a nuclear weapon when neutron-induced nuclear fission was first discovered by German scientists Otto Hahn and Fritz Strassmann in 1938 and then explained theoretically by Austrian scientists Lise Meitner and her nephew Otto Robert Frisch. These four scientists had a much more limited aim: to find out what would happen when the nucleus of a uranium atom was subjected to neutron bombardment.

Even Vannevar Bush – director of the US Office of Scientific Research and Development, which led the Manhattan Project – acknowledged this distinction in an influential report to President Harry S. Truman in 1945 that Siddhartha Mukherjee summarizes in *The Emperor of All Maladies* as follows:

> True, the [atomic] bomb was the product of Yankee 'mechanical ingenuity'. But that mechanical ingenuity stood on the

shoulders of scientific discoveries about the fundamental nature of the atom and the energy locked inside it – *research performed, notably, with no driving mandate to produce anything resembling an atomic bomb.* While the bomb might have come to life physically in Los Alamos, intellectually speaking it was the product of pre-war physics and chemistry deeply rooted in Europe.[26] [Emphasis added]

Indeed, it was Hungarian physicist Leó Szilárd, who was living in the United States when he learned about his colleagues' discovery, who had the idea that Germany might try to create an atomic bomb.[27] Szilárd shared his concerns with his colleague Albert Einstein, a Swiss physicist also living in the United States, and together they wrote to US President Franklin Roosevelt on 2 August 1939 to warn him of the risk, calling for the United States to start a nuclear programme.[28] Roosevelt took immediate action, and six years later the United States became the first country to use atomic bombs. This devastated the Japanese cities of Hiroshima and Nagasaki, killing between 35,000 and 150,000 civilians in each and ushering in the atomic age, in whose shadow we still live today.

The *physics* of neutron-induced nuclear fission is neutral: the nucleus of a uranium atom, when bombarded with neutrons, splits, forming barium as a by-product and releasing energy. This would be true regardless of whether humans had ever discovered ('found') this phenomenon. But the *intention* that drove the discovery of nuclear fission was not neutral. This was motivated by a quest for scientific knowledge, which we could argue is good or bad, depending on how we view scientific knowledge. The scientists were not doing research to facilitate the creation of a weapon. Indeed, the four scientists who discovered neutron-induced nuclear fission took the following ethical positions during the war.

- Neither Otto Hahn, who won the Nobel Prize in Chemistry in 1944 for the discovery, nor Fritz Strassmann worked on Germany's atomic bomb effort.
- Lise Meitner, who fled Germany for Sweden in 1938 after the Anschluss, refused to work on the Allies' atomic bomb project.[29]
- Otto Frisch did work on the effort to create an atomic bomb, but not for Germany; he was one of the many scientists who escaped continental Europe and worked on the United Kingdom's nuclear research programme before joining the United States's Manhattan Project in 1943.

Responsibility for the creation and use of the atomic bomb *originates* with Leó Szilárd, who first had the idea and then worked doggedly – often to the irritation of other scientists and US civil and military officials – to transform his idea into reality.[30] This responsibility extends to everyone who worked on the Manhattan Project, including Presidents Roosevelt and Truman, and everyone who has continued to create and test nuclear weapons, right up to the present day.

Does this responsibility include those who worked on the bomb *unknowingly*? Emily Strasser, whose grandfather worked on the Manhattan Project, noted that much of it was kept secret, and speculates that this was to limit the responsibility of those involved:

> If they didn't know the purpose of their work, how could they be held responsible for it? How could they question it? Secrecy was a way of protecting ideas, of keeping them out of the wrong hands. But I suspect, throughout the bomb effort at large, in addition to protecting information from enemy eyes, secrecy served to protect workers like my grandfather from the knowledge of what they were doing.[31]

This introduces a new dilemma: can we only take action about what we know for certain? Or do we have a duty to consider the possible consequences of our work even when we do not have access to the full picture?

TECHNOLOGY IS MORE THAN TOOLS

The view that technology is just a tool and what matters is what we do with it is an *instrumental* conception of technology.[32] But what do we mean by 'tool'? In what other ways can we think about technology besides being a tool? How might this open up our thinking about the 'Is technology neutral?' question and aid us in our search to pinpoint responsibility?

Tools

When we think of a tool, we probably imagine something physical, such as the bone and the atomic bomb discussed earlier, or everyday objects such as a wheel, a hammer or a compass. These are examples of a tool as something *material*, which means it is made of material substances that we experience with our senses and that exist in the physical world.

By contrast, some tools, such as software, are *immaterial*. Software exists, and we can use it and experience it being used on us, but not in the same way a hammer exists. Software has scale and complexity, for one thing, whereas a hammer is limited in scale and comparatively simple both in design and in what it is made of.

Some tools, such as a map, are *both* material and immaterial. A map can be a *physical object* that we can hold or paint on a wall, as humans did in the Lascaux cave complex in France 17,000 years ago to show star constellations and lunar cycles.[33] It may also exist as an accurate, scaled reflection of geography, such as an Ordnance Survey map.

A map can also be a *conceptual tool* used for more than just navigation of geography. For example, the map of the London Underground is not an ordnance map-like reflection of the physical geography of London, but rather a schematic diagram reflecting how the Tube works as a system to be navigated in its own right. As every Londoner knows, it can be faster to exit the Tube and walk above ground than to follow what would appear to be the optimum pathway according to the Tube map. Its value, then, is not so much in how it corresponds to the geography of London above ground but in how its connections work to get us around the city.

A map can transform how we understand the world. For example, the *Naturgemälde*, created by Alexander von Humboldt, is a map that shows the volcano Chomborazo, in today's Ecuador, with all the plants he had discovered from sea level to the summit, placing each one in the sketch where he had found them in real life.

Figure 3. Detail of Alexander von Humboldt's *Naturgemälde* (1807).

As Andrea Wulf explains, this 'translated tables of numbers into visual language' and made it possible to connect plants located at a particular altitude with other attributes found at that altitude, such as temperature, humidity, atmospheric pressure, as well as with the animals who lived there.[34] That was an innovative and powerful way to understand the information relating to this particular volcano, but the *Naturgemälde* offered much more: it was a new way of thinking about information. Once every other volcano and mountain in the world was mapped in a similar data visualization, it was possible to compare them. This led von Humboldt to come up with the concepts of vegetation zones and climate zones, as well as the 'web of life', which argues that nature is a connected, global force to be understood holistically.[35]

Maps are not neutral. They differ depending on who is creating them, from governments (especially their militaries) to companies to organizations that use open-source data and volunteers.[36]

Finally, some maps have nothing to do with geography. Rather, they visualize connections between ideas and concepts.

Technology

Technology is more than a tool. As Mark Coeckelbergh, a professor of media and communication at the University of Vienna, notes, technology can also be an infrastructure or a system, an activity or a skill, or even an idea. This both enriches and complicates the debate about whether technology is neutral.

Professor Ursula M. Franklin, the experimental physicist on team 'technology is not neutral', agrees, and in *The Real World of Technology* she helps make sense of technology by contrasting what it *is* with what it *is not*:

> Technology is not the sum of the artifacts, of the wheels and gears, of the rails and electronic transmitters. Technology is a *system*. It entails far more than its individual material

components. Technology involves organization, proce-dures, symbols, new words, equations and, most of all, a mindset... Technology is an agent of power and control... A multi-faceted entity [that] includes activities as well as a body of knowledge, structures as well as the act of structuring.[37]

We can think not only of 'technology' but of 'technologies':

- Food technologies: cooking, drying, smoking, preserving, pasteurizing, agriculture, animal husbandry.[38]
- Information and communications technologies: writing, printing, visual (art, photography, television, film), audio (music, radio/ham radio), internet and digital (websites and smartphone apps).
- Transport/distribution technologies: road, rail, air, water, space.
- Production technologies: agriculture, cottage, factories (both with humans and without).
- Hygiene technologies: water purification, soap, gels, ultra-violet lights, toilets, sanitation systems.
- Energy technologies: extractive (coal, gas, oil, mining) and renewable (nuclear, solar, hydro, wind), all of which underpin electricity.[39]

But Franklin invites us to go even further and imagine technolo-gies as more than systems of creation and production. There are also technologies of management and control, and not only of processes but of people. When, she asks, does the worker stop being in control of the process from end to end and have to rely on the knowledge and skills of others? When does a manager, or a hierarchy of managers, appear in a process?

To this we can add: when does technology start to challenge our relationship with our bodies, our family and friends, our political processes, and nature? When does it start to change our privacy, civil liberties, human rights and humanity?

When we consider technology through these different lenses, we may see the world differently. We can see democracy, authoritarianism, fascism and totalitarianism not only as political systems but also as technologies used by governments and the governed for control.[40] We can look at a device such as a smart speaker (an Alexa from Amazon, say, or a Google Nest) and see not just an AI assistant but the endpoint of a map of human labour, global resources and data protection decisions, as Professors Kate Crawford and Vladan Joler did in their 'Anatomy of an AI system' exhibit (now part of the permanent collection of the Victoria and Albert Museum in London and also available online).[41]

The grey space

In 1999 the artist David Bowie and the BBC journalist Jeremy Paxman discussed what was then a fairly new phenomenon in most people's lives: the internet.[42] They tried to work out whether it was a tool or something else:

> **David Bowie:** I think the potential of what the internet is going to do to society, both good and bad, is unimaginable. I think we're actually on the cusp of something exhilarating and terrifying!
> **Jeremy Paxman:** It's just a tool, though, isn't it?
> **David Bowie:** No, it's not. No. No, it's an alien life form!
> **Jeremy Paxman:** What do you think, I mean, when you think then about –
> **David Bowie:** Is there life on Mars? Yes. It's just landed here!
> **Jeremy Paxman:** But that – it's simply a different delivery system, though. What you're arguing about is something more profound.
> **David Bowie:** Oh yeah! I'm talking about the actual context, and the state of content is going to be so different to anything that we can really envisage in the moment, where the interplay between the user and the provider will be so in

simpatico it's going to – it's going to crush our ideas of what mediums are all about and – but it's happening in every form. It's happening in visual art. The breakthroughs of the early part of this century with people like [Marcel] Duchamp[43] who were so prescient in what they were doing and putting down the idea that the piece of work is not finished until the audience come to it and add their own interpretation and – what the piece of art is about is the grey space in the middle. That grey space in the middle is what the twenty-first century is going to be.

This interview captures how challenging it can be to comprehend the implications of the technologies we create and use – which is a necessary step towards determining responsibility.

For Paxman, the internet was just a tool, or a different delivery mechanism. He is not wrong; we can indeed argue that it is a tool and a different delivery mechanism. However, his definition was incomplete, because it failed to encompass what Bowie intuited: not only the ethical implications of the internet but the fact that it is unlike anything we had ever experienced before and that it would change (and continues to change) our ideas of media in every form.

In this Bowie was echoing Marshal McLuhan, a media theorist best known for coining the phrases 'the medium is the message' and 'global village'. As early as the 1960s, McLuhan argued that we should focus less on content than on the technologies that transmit it (e.g. television, radio and the internet) because these are more than mere delivery systems; they change *what* we do individually and collectively, and *how* we do it.[44]

So forcefully has this claim been brought home to us in recent decades that it is difficult to imagine today how we could ever go back to life without the internet. It has changed, if not everything, many things about our daily lives, creating new opportunities, new risks, new forms of connection, and new forms of segregation and isolation.

Bowie's concept of 'the grey space' is a fascinating lens through which to consider tools and technologies. Not all of them involve – or involve to the same extent – the phenomenon that sometimes occurs when creators launch their output and it is no longer in their control, much less the idea that their cycle of existence is not completed until it has been received and acted upon by other humans.

For example, the dining fork is not in the grey space. Yes, it has changed cooking and eating worldwide, over centuries, but it has also left many, if not most, other aspects of our lives untouched. While it is possible to use a fork for many things, most of us only use it for eating. We might only use a fork two or three times a day, enjoying an otherwise forkless existence, and we are not directly affected much by other people using forks.

A wheel is closer to the grey space, as it can be – and is – used in a wide range of everyday technologies and activities. As Gaia Vince observes, 'once the wheel had been invented, it became easier to imagine the potter's wheel, wagons, war chariots, wheel-barrows, gears and waterwheels' as well as all the forms of transport that use wheels today.[45] Even when we are not using a wheel ourselves, we are experiencing the effects of *other people* using wheels, whether that is by supplying us with food and other goods, by transporting us to the hospital in an ambulance (which can save our lives), or by increasing air pollution (which harms our health).

Still, the fork and the wheel, important inventions though they are, do not alter our experience of being human in the same way as information and communications technologies. For thousands of years our ability to communicate information was limited to oral transmission (storytelling, song, prayer) and visual technologies such as painting and sculpture. Then we developed writing around 5,000 years ago, papyrus scrolls (3000 BCE), parchment (1500 BCE), paper (105 CE), moveable type (1040 CE) and the printing press (1440 CE), all of which led to standardized texts, democratized information sharing, centres of learning, and continuous knowledge creation that have transformed our lives individually and collectively.[46] Computers, smartphones, the internet and AI have extended this even further by allowing us to create and share not just the printed word but audio and visual content too. They have collapsed space and time by making it possible to share such information around the world in fractions of a second.

Another example of tools and technologies that are in the grey space are those that deal with time. For thousands of years humans have tried to measure time. We developed small, portable lunar calendars that early humans made on stones, bones and antlers; astronomical observation edifices such as Stonehenge, which tracked the solstices; calendars based on celestial cycles, lunar cycles, holidays and harvests; almanacs; sundials; water clocks; astrolabes; mechanical clocks; and timelines.

However, we only achieved an accurate, standardized notion of the concept around 1840, when the invention of railways in the United Kingdom created the need to unify time so that trains could run on the same track without risk of crashing. Emily Thomas, an associate professor in philosophy at Durham University, outlines how our need to know the time led to a series of inventions that transformed our understanding of reality. First, clocks were erected on churches throughout the land. Second, people began carrying pocket watches. Third, in 1884 the Prime Meridian Conference in Washington DC led to the official adoption of a world divided into twenty-four time zones (although it took many more decades before the whole planet synchronized to this new order).[47] Finally, in 1955 the atomic clock allowed us to obtain an accurate measure of time down to the scale of nanoseconds.

Our ability to measure and standardize time not only changed how we do a task or organize a process, or even how we structure our daily lives, conceive of the past and plan for the future; it has also had profound implications for philosophy, physics, evolutionary biology, psychology, literature (writers such as

Marcel Proust, James Joyce, Virginia Woolf and William Faulkner have experimented with ways of describing how humans experience time) and computing.[48]

Even so, our understanding of time remains incomplete. Today we have three different perspectives of time: relativity (in which time is a dimension), quantum mechanics (in which time is a parameter) and thermodynamics (in which time is an irreversible arrow pointing from the past to the future).[49] For physicists – especially quantum physicists – time is an ongoing inquiry. The rest of us need not concern ourselves with this and can carry on with our day-to-day understanding of time – for now.

How does the concept of 'the grey space' help us to think about technology and, ultimately, technology ethics? One approach is to use our imagination to contemplate what would happen if we removed a tool or technology from human history.

- Remove the fork and we would miss a useful tool when preparing and eating food, but most other aspects of our existence would be unaffected.
- Remove the wheel and many more aspects of our lives would be affected, both directly and indirectly. Modern life as we know it would cease to exist. We would live hyperlocal lives; many processes would take longer; we would have a different economy because of the impact on trade; and other tools and technologies that depend on the wheel would not exist, or at least not in their current form.
- Remove information and communications technologies and our ability to understand the world and share that understanding with each other would shrink to an unfathomable degree for most of us. Innovation and collaboration on a mass scale would contract.
- Remove our ability to measure and standardize accurate time and our understanding of reality as we know it would grind to a halt.

WHERE DOES RESPONSIBILITY ENTER THE EQUATION?

Another way to think about whether tools and technologies are neutral is to examine how other species create and use them. When does a creator or user 'know' or 'understand' the consequences of a tool or technology? This can help us to pinpoint where responsibility enters the equation, which allows us to explore accountability and even liability.

In *Animal Tool Behaviour*, Robert W. Shumaker, Kristina R. Walkup and Benjamin B. Beck show that it is not just our close relatives, such as monkeys, apes and orang-utans, that use tools. Other creatures do too: elephants, polar bears, dolphins, sea otters, crows, octopi, wasps, bola spiders and more.[50] However, the authors distinguish between animals for whom using a tool is programmed behaviour, such as a hermit crab 'knowing' to carry the shell of another creature or squirrels 'knowing' to store food, and animals that use intelligence to use or make tools, such as chimpanzees making spears for hunting or wild dolphins using a sponge to flush out prey. They also question whether humans are the only animal to demonstrate symbolic tool use, which is when a tool represents something else (e.g. money), or to change an emotional state, such as when we use a special blanket or doll to comfort children.

This concept of programmed behaviour is helpful in thinking about responsibility for the use of a tool or a technology. For example, a hermit crab is genetically coded to carry around other creatures' shells. It does not have to consider a range of options and *decide* to do this. Viewed through the lens of responsibility, a hermit crab is not responsible for its actions: it is simply doing what it is programmed to do. Its use of other creatures' shells is neutral.

Is there a parallel with any tool or technology that is simply doing what it is programmed to do, such as an algorithm (a set of step-by-step rules and instructions to achieve an outcome),[51] a robot, a drone or a 'smart' system? If so, we can let those creations off the hook: responsibility for their actions lies with the humans who make them.

Things get trickier if intelligence is required to *decide* whether to use or create a tool. This is not an abstract dilemma: as of this writing, the Pentagon is debating whether to allow AI to control weapons. Should AI be allowed to decide when to kill, or should a human always be in the loop? If so, at what stage in the decision-making process does the handover from machine to human occur?

Here we must proceed with caution, recognizing that 'intelligence' is a controversial term. There are many ways to define AI – at least seventy, according to Shane Legg and Marcus Hutter,[52] and that is probably a conservative figure; as Igor Aleksander, emeritus professor of neural systems engineering at Imperial College London, has said, the definition of intelligence varies depending on whether you ask a philosopher, a psychologist or a computer scientist.[53]

I experienced this first hand in 2020 while chairing a conversation between the United Kingdom's Astronomer Royal Sir Martin Rees, the computer scientist and sex robot theorist Kate Devlin, the philosopher Hilary Lawton, and the cosmologist and theoretical physicist Laura Mersini-Houghton.[54] Each brought a fascinating, nuanced perspective to the question of whether intelligence requires *sentience* (the ability to feel, perceive and experience, with a particular emphasis on touch) and *consciousness* (to have both self-awareness and perception of our surroundings, as well as an awareness of our eventual death), and to how this might relate to being able to assess a situation and respond to it effectively. Their thinking aligned with that of Alison Adam, a professor of science, technology and society at Sheffield Hallam University, who notes that a lot of what we consider to be intelligence cannot exist without *embodiment* (our existence within our body) or culture.[55]

The more experts we consult, the more the definition of intelligence expands. For example, Max Tegmark, a professor of physics at MIT, defines intelligence as 'the ability to accomplish complex goals', while noting that it includes logic, understanding, planning, emotional knowledge, creativity, problem-solving and learning.[56] Stuart Russell, a professor of computer science at the University of California, Berkeley, agrees, and he adds reasoning, deciding, knowing and remembering into the mix.[57] Joanna Bryson, a professor of ethics and technology at the Hertie School of Governance in Berlin, points to George John Romanes's *Animal Intelligence* (1882), which defines intelligence as the capacity to do the right thing at the right time and the ability to respond to the opportunities and challenges presented by a context.[58]

This debate on what constitutes intelligence matters because some people are trying to build machines that can both think like humans *and* think differently from us. One day, machines may even surpass us in intelligence. It is an ethical minefield, so we must exercise great care as we consider *who* or *what* may be described as 'intelligent' and, by extension, capable of being held responsible for their decisions and actions – both for living things and machines.

Plants

Plants do not make tools or create technologies, but they do have sentience: they can sense their environment and adjust accordingly. Plants also go through a process of acquiring knowledge and understanding through the senses and through experience, particularly over long periods of time.

Some botanists argue that this means that plants have intelligence.[59] After all, for thousands of years plants have 'solved' problems such as drought, floods, fire, attack from other plants and animals, pollution and climate change, including by communicating among each other and by working together. Others think it is ridiculous to argue that plants have intelligence, as they do not have brains or nerves or receptors.[60] They argue that plants simply act based on programmed behaviour, not *cognition* (the process of knowing).

At stake is what we mean by intelligence. Can plants learn, decide or remember? Do they have consciousness? Or are we making a mistake in trying to apply terms designed for humans to plants? Would it be better instead to devise new terminology and definitions that would allow for 'plant intelligence' to be recognized on its own terms? Is the human way of being conscious the only way of being conscious?

For our purposes, we might venture that plants may have a form of intelligence and act to solve problems. Whether we would describe their actions as 'decisions' is highly debatable. Based on our current knowledge, it is unlikely that we would ever deem plants able to be held 'responsible' for their actions.

Non-human animals

Whether non-human animals are sentient beings that can think, feel and be described as having intelligence is a matter of considerable debate and research. The field of animal cognition explores learning, memory, perception and decision-making in animals and even extends to animal concepts, beliefs and consciousness. As Gaia Vince notes, dolphins, apes and dogs can also be described as having primitive language, enabling them to communicate well with humans.[61]

When it comes to responsibility, we tend to hold humans responsible for the actions of animals and liable for any damage they cause or harm they commit. For instance, humans who keep animals as pets or livestock can be held responsible when an animal harms or kills a human. (In general, no one is held responsible when a wild animal that is not owned by anyone harms or kills a human.) Nevertheless, consider the following thought experiment. We have already noted that a number of non-human animals can use and even modify found objects to create tools, which suggests that they are capable of making decisions, such as design choices. Could they be held responsible for creating a tool if it was ever used to harm? If animals can have rights, can they also have responsibilities?

Non-human animals invite us to think about one of the classic problems of philosophy that relates directly to technology ethics, and particularly to AI: the *mind–body problem*. French philosopher René Descartes believed that our senses, which help us to perceive our own existence and that of the world around us, are unreliable. Since they are unreliable, we must doubt them. But who is doing the doubting? Descartes's response to this question was 'Cogito ergo sum' (I think, therefore I am), by which he reasoned that the mere fact that there was an 'I' who was thinking meant that he existed.[62] For humans, our existence as thinking beings defines our essential self, Descartes argued, and our thinking mind is distinct from our unthinking body. This 'Cartesian dualism' has profoundly shaped how we think about mind as separate from body.

Yet even Descartes expressed doubts: the mind and body are distinct, but how do they relate? It was unclear in the seventeenth century how the mind worked. Although scientists and philosophers have made considerable progress on this question since then, the 'hard problem of consciousness' remains: we still cannot say with certainty what it means to be conscious.[63] This matters not only for how we think about consciousness in different living organisms, but also whether we think machines could ever have consciousness.[64]

WHAT IS IT LIKE TO BE A BAT?

In 1974 the American philosopher Thomas Nagel wrote a paper entitled 'What is it like to be a bat?' in which he argued that 'an organism has conscious mental states if and only if there is something that it is like to be that organism – something it is like for the organism'.[65] Bats, he argued, have a way of experiencing and being in the world that is unique to them, and even if we humans observe what it is like to be a bat and imagine what it is like to be a bat, we can never truly see the world from a bat's point of view or experience the world as a bat. Our human mind and a bat's mind have been formed differently since birth, not only due to the differences in our physical brains but also the fact that those brains are inside different kinds of bodies and socialized within different socializing systems.

Nagel's reflection on the subjective versus the objective point of view has implications for emerging technologies such as virtual reality (VR) and augmented reality (AR), which we can use to alter our perception of reality – when training pilots how to fly, for example, or when helping the military, emergency personnel or astronauts prepare for missions. Yet these technologies are unlikely ever to let us completely shift our point of view – in other words, even with VR and AR, we will never be able to fully experience the world as a bat.

Machines

'Can machines think?' asked British mathematician Alan Turing in 1950, and more than seventy years later we are still debating the question.[66] Some machines are sometimes described as having some version of intelligence, if they use AI. However, Kenneth Cukier, co-author of *Big Data: A Revolution that Will Transform How We Work, Live and Think*,[67] puts us on guard about the very term 'artificial intelligence':

> The term 'AI' was coined in a 1955 research grant. It didn't use the term then in vogue for the work, cybernetics (the theory of communication and control processes in animals and machines), since [the grant applicants] would have to invite MIT's Norbert Wiener, who was a pugnacious know it

all (and wrote a bestseller also titled *Cybernetics*). Hence the term 'AI' was always a misnomer.[68]

Kate Crawford agrees and argues that AI is 'neither artificial nor intelligent' but rather a dynamic system of natural resources, energy consumption, human labour in the supply chain, and data extraction from websites and devices.[69]

This is more than a question of semantics. How we answer Turing's question relates to what AI can and cannot do, and whether it can ever be held responsible for its actions or if responsibility always lies with the humans who create and use it. For instance, in September 2021 a UK court ruled that an AI system cannot be named as an inventor on UK patent applications because it does not qualify as a 'natural person' under the UK Patents Act (1977); in contrast, Australia and South Africa have ruled that an AI system can be named as an inventor.[70]

Before we go further, let us get clear on the terminology and key concepts of AI: see figure 4.[71]

As Meredith Broussard, author of *Artificial Intelligence: How Computers Misunderstand the World,* explains, there is 'general/strong/Good Old-Fashioned' AI, or what she calls 'the Hollywood kind of AI [which is] anything to do with sentient robots, consciousness inside computers, eternal life, or machines that "think" like humans'.[72] We do not have this yet. Nor do we have *superintelligence*, when machines surpass humans in intelligence – a moment computer scientists call the *singularity*.[73]

What we have is 'narrow/weak AI', which Broussard describes as 'statistics on steroids' because it 'works by analysing an existing dataset, identifying patterns and probabilities in that dataset, and [codifying] those patterns and probabilities into a computational construct called a model, which acts like a black box into which we input data and get out a numerical answer'.[74] This is not intelligence, Kate Crawford explained to *MIT Technology Review*:

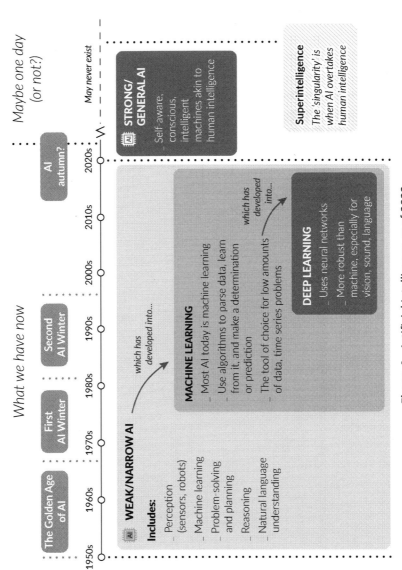

Figure 4. Artificial Intelligence as of 2022.

We've bought this idea of intelligence when in actual fact, we're just looking at forms of statistical analysis at scale that have as many problems as the data that it's given.[75]

Most AI that we use today is machine learning, although deep learning, a subset of machine learning which Broussard describes as 'machine learning on steroids', has advanced considerably in the past decade.[76] Deep learning uses neural networks, so described because, as Stuart Russell, professor of computer science at the University of California, Berkeley, explains, 'designers drew inspiration from the networks of neurons in the human brain' when creating these complex mathematical expressions.[77]

Russell notes that some researchers believe that deep learning will lead to human-level AI systems in the next few years.[78] This would make the challenge of determining responsibility even harder, because, as AI pioneer Geoff Hinton observes, 'we have very little understanding as to why deep learning works as well as it does'.[79] This is a problem because, as University of Maryland law professor Frank Pasquale explains, the term 'black box' has a dual meaning:

It can refer to a recording device, like the data monitoring systems in planes, trains and cars. Or it can mean a system whose workings are mysterious; we can observe its inputs and outputs, but we cannot tell how one becomes the other.[80]

How can we determine where responsibility lies when AI is used if we cannot fully understand how it works? We need algorithmic accountability, which means that humans should always be able to know what information was used to make a decision, how the decision was made, and what would have to change in order to achieve a different outcome.[81]

How far we have come from the first time we found a bone and realized we could use it as a tool! How far we still have to go to understand the tools and technologies that we are creating and using today, and to ensure that they help more than harm.

CONCLUSION

By now we have a sense of the debate about whether technology is neutral, as well as where some of the leading thinkers on technology ethics today position themselves, and their arguments for doing so. We have examined various tools and technologies to determine their degree of neutrality, and we have explored the idea of 'the grey space' – something fundamentally different from a 'found' tool or from tools and technologies that humans have created, ranging from the simple to the complex.

As we look to the future, what other tools and technologies might fall in the grey space? Perhaps AI, which is either 'statistics on steroids', 'more profound than fire or electricity', 'neither artificial nor intelligent' or 'could spell the end of the human race', depending on who is talking.[82] Or maybe CRISPR, a tool that will allow us to edit genes so that we can better fight viruses and design personalized medicine but that could also lead to the creation of genetically enhanced humans. Or even the Internet of Things, which might connect our world in ways that could create more synergies and efficiencies but also dependencies and risks.

We have also considered different forms of intelligence (plant, non-human animal, humans, machines) in order to ask where responsibility enters the equation. Humans have intelligence and can be held responsible for their actions, such as whether and how we create and use tools and technologies. Yet we are creating tools and technologies that could be deemed to be 'intelligent', which raises the question of who is responsible for *their* decisions and actions. Where do we draw the line?

Chapter 2

Where do we draw the line?

> There's what I call 'the creepy line', and the Google policy
> about a lot of these things is to get right up to 'the creepy
> line' but not cross it. I would argue that implanting things in
> your brain is beyond the creepy line. It is, yes. At least for the
> moment, until the technology gets better.
>
> Eric Schmidt, CEO of Google in 2010
(today an adviser to the Pentagon)

Humans have been experimenting with brain implants for years, with mixed results. Some innovations are well-established: the cochlear implant, for example, which can improve hearing. Others, such as those that allow us to control devices using only our minds, are still in the very early stages.

Such implants let us *do* things. By contrast, there is no need to implant anything in our brains in order for others to know what we are thinking. As Eric Schmidt explained in 2010, when he was CEO of Google, our smartphones and our willingness to hand over our data already give Google a good idea:

> Everyone here has a mobile device... And if you think about
> those devices – the phones and the tablets and so forth –
> they know a lot more about you. They know for example
> where you are...

If you give us information about who some of your friends are, we can probably use some of that information…

We know where you are, with your permission. We know where you've been, with your permission. We can more or less guess what you're thinking about.

Now is that over the line?[1]

Some neuroscientists certainly think so. They have called for neuro-rights to be added to the Universal Declaration of Human Rights because emerging neuro-technologies could alter what makes us human.[2]

General Michael Hayden, a former director of the Central Intelligence Agency and the US National Security Agency, prefers a box to a line. 'Give me the box you will allow me to operate in,' he said on the *Charlie Rose Show* in 2013, and 'I'm going to play to the very edges of that box… You, the American people, through your elected representatives, give me the field of play and I will play very aggressively in it.'[3] Hayden was not kidding: in 2014 he told a symposium at Johns Hopkins University, 'We kill people based on metadata' (i.e. data about data, which includes all the information associated with every email, text, social media post, digital photo and video we send).[4]

Whether we draw one line or four (i.e. a box), we are embedding our values in our technologies. Maria Giudice, a designer whose company was acquired by Facebook, explained how this works to *Fast Company* in 2019:

As designers, we have to start understanding *what the line is in terms of what should be innovated and what shouldn't be.* And that's going to be open to debate, because *it's not like the line is clear,* and you can see it with Facebook and Twitter right now. Everybody is demonizing Facebook, but for somebody who's worked at Facebook, I know how hard it is to make those nuanced decisions. *Sometimes you don't know something's wrong until it's too late.*[5] [Emphasis added]

This is not just a problem for those working in technology. Scientists know it all too well, particularly those who worked during and after World War II. As Siddhartha Mukherjee notes in *The Emperor of All Maladies*:

> The US Office of Scientific Research and Development [was] created in 1941 to 'channel American scientific ingenuity toward the invention of novel military technologies for war... New weapons needed to be manufactured, and new technologies invented to aid soldiers in the battlefield... Physicists had created sonar, radar, radio-sensing bombs, and amphibious tanks. Chemists had produced intensely efficient and lethal chemical weapons, including the famous war gases. Biologists had studied the effects of high-altitude survival and sea-water ingestion. Even mathematicians, the archbishops of the arcane, had been packed off to crack secret codes for the military.'[6]

Each of those scientists had to decide where to draw the line, and not all of them were comfortable with their decision at the time. Some changed their minds after the United States dropped atomic bombs on Hiroshima and Nagasaki. Many began to speak publicly about the moral responsibility of scientists and worked with governments to create the controls that have so far helped humanity to avert nuclear war, albeit with some very close calls and with an ever-present risk that continues to this day.

They were not alone in considering World War II to be a turning point in how we think about and apply ethics. The Allied powers created new laws after the war to deal with war crimes and crimes against humanity and then applied them during the Nuremberg Trials. The horrific revelations of the Doctors Trial, which detailed the medical experiments conducted by the Nazis on the men, women and children they had imprisoned in camps, drove medical and scientific leaders to reform medical ethics and bioethics. And then eventually, in 1948, the United Nations created the Universal Declaration of Human Rights.

Yet the Allied victors could not be complacent in victory. One ethical dilemma they faced in the aftermath of the war was what to do with the talented but tainted scientists who had worked for the Axis powers. One answer was 'Operation Paperclip', when the US government brought 1,000 German scientists over to the United States to help develop military technology to defeat the Soviet Union.[7] This meant that the US government decided *not* to hold these scientists to account for their service to Nazi Germany – despite the fact that some of them could have been tried as war criminals, so terrible were their actions. For example, Wernher von Braun was not just a Nazi and a member of the SS (*Schutzstaffel*), which was in charge of leading the 'Final Solution' (the elimination of Euro-pean Jews). He also designed the V-2 missile: a weapon that Germany used to kill an estimated 5,000 people, mostly in the United Kingdom and Belgium, and that was assembled by con-centration camp prisoners, an estimated 20,000 of whom died doing that work.[8] He later became one of the chief architects of the US space programme, helping to design the rockets that NASA used to put a satellite into orbit and to take US astro-nauts to the moon.

Was this the right decision? As far as the US government was concerned, the ends justified the means.

- The Soviet Union was now a threat to the national security of the United States and its allies.
- It was recruiting Nazi scientists to build Soviet military technology.
- If the United States declined to recruit Nazi scientists on moral grounds, it risked losing an arms race with its new enemy and thus the ability to defend US interests at home and abroad.

- Ergo, the United States decided to overlook the war records of more than a thousand Nazi scientists and recruited them to further US interests.

However, some scientists disagreed: they thought that the ends did not justify the means. Those who protested against 'Operation Paperclip' included Albert Einstein and the physicist and future Nobel laureate Hans Bethe, whose 1947 letter in the *Bulletin of the Atomic Scientists* asked:

> Was it wise, or even compatible with our moral standards, to make this bargain, in light of the fact that many of the Germans, probably the majority, were die-hard Nazis? Did the fact that the Germans might save the nation millions of dollars imply that permanent residence and citizenship could be bought? Had the war been fought to allow Nazi ideology to creep into our educational and scientific institutions by the back door? *Do we want science at any price?* [Emphasis added]

This is a question that goes to the very heart of this book: do we want technology at any price?

History shows us that where we draw the line changes over time. For example, in *Scientists at War: The Ethics of Cold War Weapons Research* Sarah Bridgers explores the generational shift that occurred in the values of scientists between World War II and the Vietnam War. During World War II many scientists were willing to build weapons and crack codes for the US government because they believed it was ethical to try to defeat the Axis powers.

By contrast, during the Vietnam War such collaboration was more controversial. This essay from a scientist writing in the *Bulletin of Atomic Scientists* in 1974 captured the mood of many:

When and where should one draw the line? Designing gas chambers for Auschwitz is clearly *over the line*, working on the Manhattan Project in the same period, clearly (to me at least) not.

What about today? Almost all of American science is supported in one way or another by the US government, in many cases through the Defense Department... To what extent should one stop cooperating with a government when one disagrees with its policies?[9] [Emphasis added]

His dilemma is immediately recognizable today, since most science and technology innovation is still supported by governments. To what extent should we stop cooperating with those governments when we disagree with their policies? To what extent is non-cooperation even possible, not only with regards to governments but also when it comes to the technology companies whose wealth and market dominance limits the options of nation states as well as individuals?

Asking *where* we should draw the line will get us only so far in tackling these questions. We also need to know *how* to draw the line and *how* we (and others) can test that we have drawn the line in the right place. This in turn means we need to know *who* draws the line and *who* decides when it has been crossed. Fortunately, there is a method for organizing and working through such questions.

HOW DO WE DRAW THE LINE (AND TEST THAT IT IS IN THE RIGHT PLACE)?

Most of us are familiar with the scientific method, in which scientists make observations, develop a hypothesis to explain what they have observed, test the hypothesis with experiments, and then publish the results so that other scientists can see if they can reproduce them. If they cannot reproduce the results – and even sometimes when they can – they offer critiques and suggestions for what to do next. It is one of the most power-ful ways that humans have created to generate, test and verify knowledge.

The scientific method emerged from another system of thinking: philosophy. This originated around the same time in India (c. 800–200 BCE), China (c. 700 BCE) and Greece (c. 600 BCE), sparking a dialogue within and across cultures and centuries that continues to this day. It remains closely linked to science, which only separated from philosophy as a formal discipline in the seventeenth century. Indeed, the word 'scientist' did not exist until 1833, when the English philosopher William Whewell coined it to describe the Scottish polymath Mary Somerville. Until then, people we would call scientists today were called 'natural philosophers'.

As with the scientific method, philosophy allows us to apply systemic rigour to our attempts to answer questions. It is com-prised of six main branches: metaphysics, epistemology, politi-cal philosophy, logic, aesthetics and ethics. While it is possible to work with these branches separately, they can also interrelate and overlap. For instance, our focus in this book is technology ethics, but ethics does not exist in a vacuum; it exists in a phil-osophical framework. We need at least a basic grasp of this framework because, as the British philosopher Julian Baggini argues:

> Understanding the philosophical framework of a people is like understanding the software their minds work on. If you don't know their software there will always be this gap in terms of understanding conversation.[10]

In this chapter we will think of philosophy as the software our minds work on, or, for those who prefer a more tangible meta-phor, as a Swiss army knife: a tool containing six components tools that we will use to address questions of technology ethics. Since philosophy is a vast subject and each component is a topic in its own right, our goal here is a minimum viable product: that is, an understanding of philosophy that we can get started with, while bearing in mind that there is so much more we would need to do to make it full-featured. What follows is a short definition of each branch of philosophy accompanied by examples of how each applies to technology.

Metaphysics: what is reality?

Metaphysics is concerned with the study of the nature, structure and origins of the universe – in other words, reality. The early natural philosophers asked, 'What is the universe made of?' and they answered with what we would today call the natural sciences, e.g. astronomy, physics, chemistry, biology and botany. Scientists still debate metaphysical questions today, and even work with their philosophy colleagues in asking metaphysical questions such as, 'What is the relationship between our mind and our body?' and 'What is consciousness?'

As applied to technology
One of the biggest metaphysical challenges today is the polluting of the information ecosystem – something that means that people do not have a shared sense of reality. In a post-truth world, 'facts are irrelevant, narrative is everything'.[11] This can have dangerous consequences, such as the undermining of confidence in elections, incitement to violence, or the belief that a virus is not real (even as it kills and makes others ill all around us).

'What does it look like if we don't have a shared sense of reality?' mused Claire Wardle – executive director of First Draft, a group that researches and combats disinformation – in an interview with *The Guardian* in 2021. 'We've seen more conspiracy theories moving mainstream. There's an increasing number of people who do not believe in the critical infrastructure of a society. Where does that end?'[12]

Another example of how metaphysics applies to technology is Alan Turing's question, 'Can machines think?' This forces us to ask about the reality of thinking: 'What is thinking? What is intelligence? What is consciousness? What does it mean to have a mind?'[13]

Metaphysics can seem abstract (e.g. 'What is a soul?'), but we can use it practically too. David Silver, who helped DeepMind create a programme that taught itself to play the boardgame Go and defeat a grandmaster, uses metaphysics to frame the challenge of how to create an AI as powerful as a human brain:

> The first step in taking that journey is to try to understand what it even means to achieve intelligence. What problem are we trying to solve in solving intelligence?[14]

Facebook Reality Labs is also using applied metaphysics when it offers us virtual reality, which 'fully immerses people in 3D virtual environments' using headsets, and augmented reality, which 'takes computer-generated images and overlays them onto our view of the world' using the camera on our smart devices.[15]

And so is the US Army, which in 2021 awarded Microsoft a contract worth $21.88 billion over ten years to build 120,000 augmented reality headsets.[16] CNBC reported that this will be a customized version of its HoloLens, which allows users to 'see holograms over their actual environments and interact using hand and voice gestures'.[17] A prototype built in 2019 showed the user a map, a compass, thermal imaging (to spot people in the dark) and the aim for a weapon.[18]

Apple use a LiDAR (light detection and ranging) scanner in the iPhone 12 Pro, iPhone Pro Max and iPad Pro to scan our environment and create a three-dimensional model of it.[19] This improves the functioning of the devices' cameras in low lighting, measures distances in a closed space (such as a room) and measures a person's height or the dimensions of an object (such as a table) – all of which enhances augmented reality.

We use applied metaphysics whenever we query a dataset or consider if the output of a model or an algorithm bears any resemblance to our understanding of reality. For instance, in 2018 whistle-blowers at Amazon revealed that for five years the company had been using an AI recruiting tool that was biased against women applicants.[20] The program reviewed applicants' curricula vitae in order to select candidates to invite to interview, but it was trained on a dataset of CVs submitted to the company over a ten-year period, and thus it contained a bias: the majority of applicants who had applied to the company over that period were men. On the one hand, this was an accurate reflection of reality: many more men work in the technology sector than women. On the other hand, it should have raised a flag because it does not reflect the reality that women make up half the population.

This (mis)perception of reality created a problem, because the algorithm drew a conclusion from the dataset without being instructed to do so: it inferred that men were the best candidates to invite to interview and penalized CVs that included the word 'women'. When the developers discovered this error they edited the programs so that female candidates would no longer be penalized if, for example, they had been educated at a women's college or were captain of the women's tennis team. But still the algorithm persisted in discriminating on the basis of sex, because the program had taught itself to identify female applicants from the data even when information that explicitly identified them as female was removed. The women applicants would have had no way of knowing this, and neither would human resources or anyone else involved in talent acquisition.

Applied metaphysics could have raised the alarm:

Question: Does it make sense that all the candidates that this program is inviting to interview are men? Does this reflect the reality of the population?

Answer: Well, men have historically applied to jobs at Amazon in greater numbers than women.

Question: Yes, but what does that really tell us? Is that because men truly are the best candidates for the jobs we offer? Or could there be *any other reason* why women do not apply to Amazon in equal numbers to men or are rejected more by our recruitment process?

From there, Amazon could have gone on the hunt for data to test its understanding of reality, asking more easily answered questions, such as the following.

- What is our track record on hiring when it comes to diversity, equity and inclusion?
- From where have we been sourcing new talent?
- Are we overlooking any talent pools?
- What kind of language are we using in our job postings? Is any of it attracting some candidates and turning off others?
- Is any of this reflected in the AI recruiting tool we have been using for five years? Have we had it audited by an external party?

Of course, all of this assumes that Amazon might have wanted to know why its AI was selecting men and rejecting women.[21] But who at Amazon should have wanted to know? Developers look to observability and traceability in their software systems to find errors, not to establish whether the data is representative of 'reality'. That is the job of the data scientists. Does this point to the need for data scientists in software development or for developers to add this ability to their skillset?[22] Alas, we may never know what Amazon learned from this because it shut down its sexist AI recruiting tool following the whistle-blowers' action and declined to answer questions from the media. This question of what can be known leads us to the next branch of philosophy: epistemology.

Epistemology: how can we know?

Epistemology explores knowledge, the methods for acquiring it, and its limitations.[23] This includes learning from experience, including using our senses to make observations, and reasoning, because we cannot rely on our senses alone.[24] It also relates to questions of authority, such as what is a source of knowledge? Who is an authority on knowledge? How is this authority conferred? Finally, it relates to questions of information integrity: is this data, evidence or fact solid?

As applied to technology
To answer Turing's metaphysical 'Can machines think?' question, we must also ask epistemological questions. How can we know if machines think, for example? Who or what are our sources of knowledge on this question?

This gets right to the heart of AI today. On the one hand, AI is a tool which can do developers' job better by running millions of experiments and averaging the results. Good AI should be traceable and able to be modelled.[25]

On the other hand, AI is also dominated by technology giants whose datasets and code are their intellectual property, so they do not always want to make them available for scrutiny.[26] How, then, can we check to see if their results can be reproduced (a key criterion of the scientific method) let alone test for bias, accuracy and other aspects of algorithmic transparency and accountability?[27] For instance, do we want humans to have the right to an explanation when AI makes decisions about whether to approve a loan or proceed with a job application, as Article 22 of the EU's data protection laws guarantees (and which the post-Brexit United Kingdom is looking to remove)?[28]

Even if all datasets and algorithms were made available, it would still not always be possible for researchers to know how an AI reached a conclusion, particularly where 'deep learning' is involved. This is known as a 'black box', which Frank Pasquale described as 'a system whose workings are mysterious; we can observe its inputs and outputs, but we cannot tell how one becomes the other.'[29]

Another example relates more directly to freedom of information versus secrecy or censorship. Sometimes researchers and technology workers have the information they need when confronted with an ethical decision. This was the case in 2018, for example, when some Google employees protested against working on a project to build autonomous weapons for the US Department of Defense (Project Maven), and others protested against the company's work to build a search engine that allowed China to censor online searches (Project Dragonfly).[30]

At other times, though, technology workers are prevented from knowing the full implications of their work. In 2018 some Silicon Valley workers re-examined the implications of the fact that their start-ups were being financed in part by VisionFund, an investment partnership between Japan's SoftBank and the sovereign wealth fund of Saudi Arabia, whose regime ordered the murder of the *Washington Post* journalist Jamal Khashoggi. This led to an article in the *New York Times* with the headline 'Start-ups ask, "Are we making money for Saudi Arabia?"' VisionFund's connection to Saudi Arabia was well known, of course, but since venture capital firms rarely provide information about their sources of funding, it can be difficult for founders – or anyone else – to follow the money.[31] This makes it challenging for technology workers to know if they are creating (un)ethical technology.

Other examples of epistemology as applied to technology are the company missions of Google ('to organize the world's information and make it universally accessible and useful') and Wikipedia ('to benefit readers by acting as a free, widely accessible encyclopedia').[32] Both missions trace their origins back to the French Enlightenment, when a group of philosophers began working in 1745 to translate an English dictionary and ended up forty years later with the *Encyclopédie*: a bestseller of seventeen volumes of text and eleven volumes of illustrations cataloguing all known knowledge of science, technology, travel and the arts.[33]

Then, as now, questions over who creates knowledge and what topics are worthy of being codified were controversial. The *Encyclopédie* was written almost entirely by men, and the same is true of Wikipedia, one of the most visited websites on the internet: as of 2019, women made up only 15–20 per cent of Wikipedia's contributors.[34] Is it cause or correlation that explains the relationship between Wikipedia's mainly male contributors and the fact that (as of 2019) articles about men exceeded those about women by about four to one?[35] For Katherine Maher, who until February 2021 headed the Wikimedia Foundation that runs Wikipedia, the answer is that the relationship is causal. As she told *The Guardian*: 'If we have more women editing Wikipedia, do I expect more articles about women scientists and novelists? Absolutely.'[36]

Epistemology also grapples with how we acquire knowledge in the face of censorship – from religious authorities, the government and the private sector, for example. Journalists, historians, researchers and others who organize and safeguard knowledge (such as librarians, archivists, museum curators, publishers of reference resources) have long been on the frontline of epistemology. They find existing information, create new sources of information, and play a role in determining what we remember – and what we forget.

Finally, epistemology relates to how we distinguish knowledge from *misinformation*, which is 'incorrect or misleading information', and *disinformation*, which is 'false information deliberately and often covertly spread in order to influence public opinion or obscure the truth'.[37] Both are harmful to the information ecosystem, but disinformation is especially sinister because it is *intended* to harm, e.g. by seeking to persuade people not to get a vaccine that would protect them and their community, or by sowing distrust in free and fair elections.[38] Both are powerful forces because we humans sometimes choose our 'facts' and sources of knowledge in alignment with our personal and political beliefs, whether consciously or unconsciously.[39] Social media intensifies this because of the scale and speed with which it disseminates information. For instance, a study by researchers at the Massachusetts Institute of Technology in 2018 found that falsehoods almost always go viral faster, and travel further, on Twitter than truth does.[40]

For many of us, knowledge is a matter of opinion rather than something that must be established, tested and verified. It is also a question of trust. Some people claim to distrust experts, such as Michael Gove, who in 2016, while he was UK Justice Secretary, said: 'I think the people in this country have had enough of experts.'[41] The risk here is an inability – or even a refusal – to recognize the difference between having an opinion and having expertise.

- We all have opinions.
- We do not all have expertise.
- Even experts do not have expertise on all topics. Those who use their expertise in one topic to claim expertise in another risk 'epistemic trespassing' – and public smackdowns by real experts.[42]

The solution to this is not simple. Values of humility, courtesy and respect are a good place to start. Sometimes we really can agree to disagree. However, that may not be possible when topics of disagreement and distortion have life or death consequences, or when they curtail our rights, or when they lead to violence, hatred or distrust in democracy.

Some social media platforms and traditional media outlets have recently taken notice of these problems and taken steps to address the issue.[43] For instance, in 2020 Twitter began fact-checking US President Donald Trump's account, adding commentary whenever his tweets contradicted reality, such as when he lost the US presidential election to Joe Biden.

Figure 5. Twitter deployed epistemology (the authoritative knowledge of the election officials of fifty US states) to defend an attack on metaphysics (Trump's repeated attempts to deny the reality of Joe Biden's victory in the 2020 US election).

Yet it is far from clear how to remedy what the World Health Organization calls an 'infodemic': an overabundance of information and the rapid spread of misleading or fabricated news, images and videos.[44] Science writer Philip Ball suggests that we need a public health strategy to fight it:

> Some have suggested the idea of 'inoculating' populations in advance with reliable information, so that false ideas can never get a foothold (although that is surely much harder now there is such widespread distrust of 'elites'). We need agreed and enforceable standards and regulations for social media. We need diagnostic tools to rapidly identify and isolate 'super-spreaders', and 'virologists' of misinformation who can find and attack its weak spots. And we need to understand why different people have different levels of immunity and susceptibility to bad ideas – and to recognise that understanding misinformation, like disease, is in many respects an inescapably sociopolitical affair.[45]

However, as with global public health, one challenge with fighting the infodemic is that it is a global phenomenon, and one that is increasingly weaponized *within* nations as well as *between* them.[46] To formulate an effective mitigation strategy, we must first the following questions.

- Who benefits from misinformation or disinformation?
- Who is harmed?
- Who gains when we live in a world that is not only post-truth but post-trust?

To help us answer these questions, we can use logic.

Logic: how do we know what we know?

We use logic to determine whether an argument is sound or if a hypothesis is false. At the most basic level, there are two main categories of argument.

- *Deductive arguments* start with general conditions and move to specific conclusions. The classic example is Aristotle's syllogism: 'All men are mortal. Socrates is a man. Therefore Socrates is mortal.'
- *Inductive arguments* start from specific conditions and move to a general conclusion: for instance, 'All the swans I have seen in my life are white. Therefore all swans are white.' This is a 'weak' inductive argument, because black swans exist. We can 'strengthen' the argument by qualifying it: 'There-fore most swans are probably white.'[47]

Already we can see the dangers of inductive arguments: even though the premise may be true (all the swans we have ever seen really have been white), the conclusion may be false (despite our personal experience, black swans do exist, there-fore not all swans are white). Nevertheless, inductive arguments are necessary to test hypotheses so they remain a valuable tool. We can ask further questions to avoid inductive logic pitfalls. For example, how many swans have we seen? How representative is our sample size? Are we only looking where white swans are and not looking where black swans may be?

Logic can also be expressed in pure mathematical terms or in simple sentences, by mathematicians, philosophers, computer programmers and the rest of us, any time we need to reason through a problem or make a decision.[48] We can learn from sound logic as well as unsound logic, as this section explores.

As applied to technology
In *Daemonologie*, published in 1597, King James VI of Scotland (later also King James I of England) describes two tests to determine whether a person is a witch. The first was to find the devil's mark on the accused's body. The second was to strip the accused naked, tie their thumbs to their toes, and put them in a river. If they floated, they were a witch. If they were innocent, they sank.[49] The logic of the swimming test was based on a belief that God had ordered any water to reject anyone who had committed such monstrous impiety.[50] Today we wonder how anyone could ever have thought this was a good test: the flaws in the logic are so easy to spot! Yet before we start feeling too smug, we may wish to consider that one day people may look back at us and be baffled by the logic of tests currently used to determine if we are a robot.

To answer the question 'Can machines think?' Alan Turing proposed the Turing test, which Oxford University computer scientist Michael Wooldridge summarizes as follows:

> You are interacting via a computer keyboard and screen with something that is either another person or a computer program. The interaction is in the form of text – questions and answers. Your task is to determine whether the thing being interrogated is in fact a person. Now suppose, after some time, you cannot tell whether the thing is a person or program. Then you should accept that the thing being interrogated has human-like intelligence.[51]

This will sound familiar to anyone who has been online: most of us have been tested when we signed into our online accounts, made purchases or posted reviews in order to ensure that we were humans and not spam bots, which take people's email addresses, commit fraud and post spam-filled messages on public forums.[52] The tests are called CAPTCHA, which stands for Completely Automated Public Turing test to tell Computers and Humans Apart. Figure 6 shows two examples of a CAPTCHA asking us to click a checkbox to say we are not a robot.

Figure 6. Are you or are you not a robot?

Figure 7 (overleaf) shows another common CAPTCHA: one that requires us to select objects in a grid of photos.

In 2017 Google claimed that its reCAPTCHA test was so good that it could run in the background of websites, invisible to users unless they were flagged as suspicious, at which point they would be tested.[53] That is because logic underpins CAPTCHA tools. For years millions of us have been tested every day, creating enormous datasets on which algorithms have been trained to recognize what is likely to be human behaviour and what is likely to be a bot. That is why Aaron Malenfant, the engineering lead on Google's CAPTCHA team, told *The Verge* in 2019 that 'five to ten years from now, CAPTCHA challenges likely won't be viable at all. Instead, much of the web will have a constant, secret Turing test running in the background.'[54] But this poses

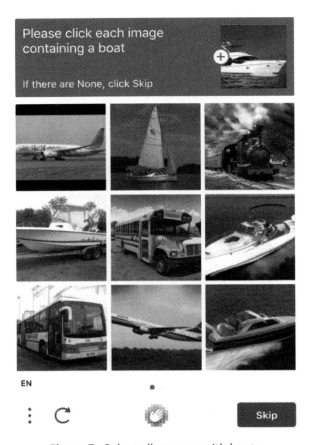

Figure 7. Select all squares with boats.

questions. Should we have the right to know that we are being subjected to a constant, secret Turing test? How does Google's invisible Turing test square with the AI ethics values of transparency, explainability and accountability? What happens if we are unable to distinguish humans from bots online?

Another possibility is that CAPTCHA tests will continue in the open but be based on a logic whereby the only way to pass the test is to get the answer wrong – because machines will perform better on the tests than humans![55]

Figure 8. 'Machine learning CAPTCHA' by
XKCD.com (used with permission).

Whatever the fate of CAPTCHA, this discussion is a useful introduction to one of the biggest challenges in technology. At the
heart of testing whether you are a witch, a robot, or a human
lies a fundamental problem: how can we test that you are *you*?

Gus Hosein, Edgar Whitley and other identity scholars frame
this using three questions.[56]

- Identification: what do we know about you?
- Verification: how do we know it is you?
- Authorization: what do we allow you to do?

We will explore these questions in depth when we come to look
at facial recognition technology and digital health technologies,
as identity is at the very core of how these technologies are
changing the balance of power between citizens and the state,

and between consumers and companies. It is to this question of power that we now turn.

Political philosophy: what is the nature of power?

Political philosophy explores the relationship between the individual and society. It examines concepts such as power, authority, legitimacy and freedom. It asks questions such as what sort of society do we want to live in? How should we organize it? Who should rule? What are people's rights and responsibilities? What does it mean to be free?

As applied to technology
Some technology companies are effectively political actors at the national, or even international, level. Meta (which owns Facebook, Instagram and WhatsApp), Alphabet (the parent company of Google and YouTube), Twitter, Apple, Microsoft and Amazon are all in this category. So is China's Huawei, a world leader in mobile and 5G telecommunications technology that has been shut out of many liberal democracies because of its ties to the Chinese Communist Party and its role in the Chinese government's persecution of the Uyghurs.[57] The power of these companies is derived from a combination of factors, such as their wealth, the number of customers they have and their market dominance (to the point that some are arguably monopolies). Other factors are directly linked to their existence as technology companies, such as their ability to help shape political discourse, sometimes with deadly consequences.

For example, Matthew Prince, the CEO of Cloudflare, which protects websites from cyberattacks, has twice been tested on his company's position of being 'content neutral'. In 2017, after the white nationalist rally in Charlottesville, Virginia, he had to decide whether to stop protecting the Daily Stormer website, a neo-Nazi hate site; and in 2019 he had to decide whether to take down 8chan, an anonymous message board that had been used by a man accused of carrying out the El Paso massacre.[58] In both cases, Prince abandoned his position of neutrality and stopped protecting the sites, but he did not relish his power: 'I woke up in a bad mood and decided someone shouldn't be allowed on the internet. No one should have that power.'[59]

Who *should* have that power?

Figure 9. 'Free speech' by XKCD.com (used with permission).

This is not the only question that most lawmakers worldwide have failed to answer in response to the challenges posed by technology companies. Thanks to their data-gathering and analysis practices, these companies are challenging our understanding of power relations, particularly with regards to what constitutes privacy, civil liberties and human rights. Those that specialize in internet search and social media have considerable power over what is knowable – and with what emphasis – and what is not knowable. For example, in January 2021 Google admitted to hiding some Australian news sites from search results while it was in a dispute with the Australian government. The government ultimately went on to pass legislation requiring Google and Facebook to agree to pay news providers for content.[60]

Finally, the 'Who controls our data?' question also falls within political philosophy, as it relates to the privacy, agency, civil liberties and human rights of not only ourselves but of everyone in our social network. The very question is provocative in its framing, for it requires us to decide whether data control is a question of *ownership* or one of *rights*. Not all countries or people share the same view on this point. The fact that laws differ on this complicates our discussion and also makes cross-border data transactions more difficult.[61]

Aesthetics: what is experience?

Aesthetics is usually understood to be concerned with questions of beauty. For instance, does beauty exist objectively, or is beauty in the eye of the beholder? Why are things beautiful? Is beauty only a matter for our senses, or does intellectual, moral or functional beauty exist? If those other forms of beauty exist, does the beauty of the senses help or hinder them? For this reason, aesthetics is associated closely with the study of the arts. However, Julian Baggini notes that it was only in the nineteenth century that aesthetics was commonly associated with beauty; before then, the word aesthetic meant 'relating to felt experience' and 'centred on the experiential'.[62] This involves our senses and perception, and some philosophies – such as those of India, China and Japan – offer perspectives on aesthetics that go well beyond art and beauty.

This can sound pedantic and formal, but we have many aesthetic experiences in our everyday lives – as we prepare and eat our meals, for example, or as we select our fashion style and create our look or design our interiors and surroundings. We have them alone – such as when we are in nature and when we meditate, pray, listen to music, read, create something or soak in a hot bath – and we have them with others, when we engage in dialogue, dance, listen to live music, watch a play or a film, have sex or cheer on our sports team. We seek out aesthetic experiences when we try new things, travel or move somewhere new. All of this expands and enriches our experience of being human, which is increasingly taking place in both the physical and digital worlds.

As applied to technology
'In designing tools, we are designing ways of being,' wrote Terry Winograd and Fernando Flores in *Understanding Computers and Cognition: A New Foundation for Design.*[63]

Aesthetics is a central concern when it comes to the creation and use of technology and tools. This includes form and function as well as user interface (UI) and user experience (UX).[64] These things do more than determine the attractiveness of a tool or technology: they are critical to 'stickiness', which hooks users and creates and maintains brand loyalty.[65] Done well, we will often fail to even notice why a tool or technology works or is pleasing to use – we simply integrate it into our experience. Done badly, we experience a tool or technology as anything from unsatisfying to infuriating.

Aesthetics as user experience manifests in different ways. It might mean making digital technologies more accessible and inclusive. For example, the UK Government Digital Service suggests thinking about the design needs of users with low vision, users of screen readers, users who are deaf or hard of hearing, users with physical or motor disabilities, users with dyslexia, and users on the autistic spectrum.[66] It could also mean finding ways to verify a user's identity that recognize the difference between sex and gender.[67]

The consequences of design that fails to uphold the values of accessibility and inclusivity can be more than just frustration and exclusion for the user: it can be a matter of life and death, and it can be illegal. For example, during the Covid-19 pandemic, WebAIM and the US non-profit Kaiser Health News discovered 'technological accessibility barriers on all but 13 of 94 US state webpages that included information about the vaccine, lists of providers and sign-up forms', which violates the Americans with Disabilities Act and the Rehabilitation Act of 1973.[68] As Cyndi Rowland, the executive director of WebAIM, told *NBC News*:

> It is, I think, a sad reflection of our times that accessibility is still not part of what is top of mind for developers. We've had web accessibility guidelines and standards for 20 years. This has been a real rough year for folks with the kinds of disabilities that affect computer and internet use.[69]

Aesthetics can also inspire how we conceive of a solution to a problem. For example, in the very early days of the pandemic, Audrey Tang, Taiwan's digital minister, commissioned a national tracking system for mask supplies as the coronavirus began to spread around the world. As she explained to *WIRED*, 'the significance of the mask map portal was its function as a space for *others* to participate in'.[70] She went on to cite the *Dao De Jing*: 'Hollowed out, | clay makes a pot. | Where the pot's not | is where it's useful.'

Aesthetics might guide how data scientists or journalists create a data visualization to help an audience understand cause and effect, decision trees and path dependencies. It shapes urban design as our cities become 'smart' – equipped with sensors to monitor pollution and traffic, and with cameras to surveil us – inviting us to think about who we design cities for, who we exclude, and how to make them more resilient and climate friendly.

In *Design Justice*, Sasha Costanza-Chock discusses how aesthetics relates to political philosophy, since the way objects and systems are designed can encode, reinforce or challenge power, for example, through design settings based on a default gender, race, class or language.[71] Value-sensitive design asks who should be involved and what values are implicated in design in order to create tools and technologies that are more inclusive and accessible.[72]

Aesthetics can also be about curiosity, fun and joy, which can translate into creating and using tools and technologies that we love. As a college student, Steve Jobs, the founder of Apple and co-founder of Pixar Animation Studios, took a calligraphy class purely for fun and became passionate about typefaces and spacing. He later explained the influence this had on his work:

> None of this had even a hope of any practical application in my life. But 10 years later, when we were designing the first Macintosh computer, it all came back to me. And we designed it all into the Mac. It was the first computer with beautiful typography. If I had never dropped in on that single course in college, the Mac would have never had multiple typefaces or proportionally spaced fonts. And since Windows just copied the Mac, it's likely that no personal computer would have them. If I had never dropped out, I would have never dropped in on this calligraphy class, and personal computers might not have the wonderful typography that they do.[73]

Jobs's collaboration with designer Sir Jony Ive and with Susan Kare, who created many of the icons and typefaces used by Apple (as well as Facebook, IBM, Microsoft, PayPal, Pinterest and others), resulted in products that are pleasing to the senses, intuitive to use and instantly recognizable.[74]

Did this aesthetic derive at least in part from Jobs's long-standing interest in Hinduism and the practice of Zen Buddhism and their emphasis on developing intuition? He certainly thought so, telling his biographer Walter Isaacson, 'I have always found Buddhism – Japanese Zen Buddhism in particular– to be aesthetically sublime', and acknowledging its influence on his love of simplicity in design.[75]

Aesthetics, while a branch of philosophy in its own right, is also about more than just experience and beauty. It can be a *value*, one appreciated by some people more than others. This brings us to ethics, the branch of philosophy that addresses questions of values and is key to our focus on technology ethics.

Ethics: what are our values?

Ethics is concerned with right and wrong, with how we should live and with what constitutes a good life, as well as with where our values come from and on what basis we hold those values. It is the focus of this book's central question: how can we create and use technology to maximize benefits and minimize harm?

This question derives from utilitarianism, which was developed by the British philosophers Frances Hutcheson, Jeremy Bentham and John Stuart Mill, who argued that we should try to maximize the happiness of the greatest possible number of people.[76]

As a lens for studying technology ethics, utilitarianism has pros and cons[77]. It overlaps neatly with the 'Does it scale?' question, which in this context means: 'Can this tool or technology be used by the most people?' Technologies and tools that scale are the opposite of those that are described as niche: that is, ones that are used by fewer people, usually because they are more expensive, offering the possibility of something bespoke, tailored or luxury. While there is always money to be made in niche products and services, 'scale' is of particular interest to technologists, those who invest in their creations, and those who seek to use those creations to solve problems that affect a large number of people. For investors, scale is where the most money can be made. For governments, scale offers the potential to reach the most citizens, residents and visitors, for the minimum cost. For consumers and citizens, scale often creates cheaper goods and services.

However, for all its strengths as a lens through which to examine technology, utilitarianism has some serious drawbacks.

First, it offers little protection to minority populations, leaving them at risk of the tyranny of the majority.

Second, it veers uncomfortably close to the argument that 'the ends justify the means', which was advanced by the Italian politician and diplomat Niccolo Machiavelli in *The Prince* and which implies that unethical means can achieve ethical goals, thereby allowing for harm along the way.

Yet not everyone accepts the premise of utilitarianism, as it prioritizes outcomes above all else. Some philosophers argue that it is not possible to use unethical means to achieve ethical ends. For example, the Greek philosopher Aristotle and the German philosopher Immanuel Kant argue that it is intention, not outcome, that defines whether an action is ethical.[78] By contrast, the Dutch–British philosopher Bernard Mandeville argues in *The Fable of the Bees* that a focus on virtuous actions is itself misguided because some private vices can have a public benefit – particularly those that provide employment and can be taxed.[79]

Finally, utilitarianism fails to engage with an important design challenge for any technologist: how to build tools and technologies that can scale while also allowing for customization and personalization.

For all its flaws, utilitarianism is still a powerful lens through which to study technology ethics. It can be applied to so many of the problems that confront humanity, and its very weaknesses remind us of the importance of the individual, of minority groups and of concepts such as universal human rights. Even if we wanted to, we could not ignore it because its inventor, Jeremy Bentham, also developed the concept of the panopticon, a method of mass surveillance that we will explore in the chapters ahead.

As applied to technology
Technology ethics is an example of *applied ethics*, or ethics that we put into practice. We are 'doing' technology ethics whenever we study, research or work in data ethics, AI ethics, engineering ethics, computer ethics, design ethics, ethics of robotics, ethics of human–robot interaction, ethics of drone technology, ethics of developing technologies, ethics of algorithms, and so on.

Various academic disciplines have been working on technology ethics for a long time, and they offer a rich body of knowledge on which we can draw.

- *Science and Technology Studies (STS)* explores the relationship between science/technology and society, politics and culture.
- *Digital Humanities* applies computer-based technology to the humanities to create digital tools, methodologies, archives and databases researching and analysing content and data.
- *Human–Computer Interaction (HCI)* focuses on the interfaces between humans and computers and therefore sits at the intersection of many disciplines, such as design, the behavioural sciences, computer science, and many more.

Efforts are underway to build bridges between academia and the so-called real world to develop methodologies that we can use to embed ethics in the creation of tools and technologies. This underscores how technology ethics is not simply a checklist that we can delegate to one person or team to do once, or even a few times a year, and then forget about. It is a mindset. It is a conversation that is interdisciplinary, dynamic and iterative – rather like technological innovation itself.

Who should adopt this mindset and have this conversation?

Technology ethics is not only the responsibility of the legal, policy or public relations teams, or of software and product developers. It concerns everyone, from the moment someone first has an idea to the moment a product or service is brought to market or otherwise put into action.

1. *Someone has an idea for a new product or service, or to enhance one that already exists.* Not every idea makes it to market; most do not. This is the first stage where ethics comes into play, because the person (or people) with the original idea must consider whether to take it forward. Depending on various factors, they may decide to abandon it.

2. *Someone finances the idea.* This may involve the person (or people) who had the idea using savings or getting a loan, but they could also seek investment from their employer or funding from investors (venture capitalists, private equity, banks, the government) and, at a later stage, shareholders. All of these people have considerable influence over whether ethics is a consideration in the creation of tools and technology.

3. *Someone transforms the idea into reality.* This could involve the person or a group of people who originally had the idea, such as a start-up, perhaps, or a more established public or private company or a government. It could involve leaders – such as the executive-level managers within a company (the C-suite) and board – legal, policy, HR, marketing, sales, customer service and others, as well as the product development and technical teams. It may also involve partners (e.g. suppliers, contractors and other third parties) without whom the idea could not be brought to market. Organizational culture matters because we need to know if anyone can stop a product or service from being launched if there is an ethical concern at this stage. If not, ethics is a risk in this organization.

4. *Someone makes, builds and creates something based on choices about hardware, software and data and design.* Considerations include the physical design of a product, user experience, user interface, power relationships, and impact on sustainability, the environment and climate.

5. *Someone gives feedback.* While this also happens before the final stage, we can put it here to capture the 'broad release' phase, which occurs when a product or service is released into the market.

In reality, feedback in modern start-ups is often sought early, and then stages 4 and 5 in this list are repeated again and again in a loop. This is the concept of 'human-centred design' on which many start-ups run.[80] Initial feedback is also sought from employees, close friends, investors, focus groups, etc., although this early feedback is not representative of the broad release group of customers and/or clients, next-stage investors, share-holders, the media, regulators, lawmakers and governments. Furthermore, broad release feedback may also target investors, particularly on ethical issues, as well as the creators.

How tidy it looks on the page! It's a wonder anything ever goes wrong.

Of course, there is a world of difference between how tech-nology ethics might work in theory and what actually happens in reality. Ethics is not always a consideration. More often, the focus is on viability or survivability. So when *should* founding teams start thinking about ethics: after they have found some traction for their idea or from the outset?

There is a parallel here with how some digital services approach accessibility. Too often they fail to build services that, from the outset, account for the accessibility requirements of screen readers and other accessibility tools, which is often a legal requirement. Instead, they try to build this in later, which is often more expensive because it requires re-architecting things. Teams that build using accessibility right from the start often do it for the same cost, and it is more future-proof. Similarly, those who incorporate technology ethics into their thinking from the outset may enjoy benefits and advantages compared with those who do not. But be warned: technology ethics is not for the faint of heart. Any misalignment of values, words or actions, or any inconsistency or hypocrisy, is a fair target for the media, researchers, employees (past, present and future), lawyers, law-makers and regulators.

This means that technology ethics should be on the risk radar of anyone who funds the creation and deployment of technology.[81] While there will always be investors who are happy to profit from surveillance, censorship and other controversial activities, others consider technology ethics to be part of their due diligence. In a 2019 article for the *Washington Post*, Katherine Boyle, a venture capitalist at General Catalyst, poses the following example questions.

- 'Which authoritarian regimes is it okay to make rich by letting them invest in your company?'
- 'Is it worse to pay teens for their mobile activity data or to sell them lollipop-flavored nicotine?'
- 'If you refuse to work with the [US] Department of Defense, is it okay to work with Beijing?'[82]

Sometimes it is easy to spot technology ethics, such as lethal autonomous weapons (drones, tanks, automated machine guns and robots) that can operate and kill without the need for human involvement.

At other times, more imagination might be required: for example, when considering the dangers of non-military drones that can be purchased for commercial or personal use and operated without a licence or registration. These drones are fun, useful and capable of bringing one of the world's busiest airports to a halt for three days, as occurred at London Gatwick in December 2018. This was the moment when the world learned that no UK airports have anti-drone capabilities.[83] How did no one foresee this risk when this product was allowed to be sold throughout the United Kingdom, priced so affordably that our parks are full of drone enthusiasts alongside people flying kites?

Even greater creativity may be needed to think through the risks of social media platforms. In 2004 Mark Zuckerberg created Facebook so that male undergraduates at Harvard could rate the attractiveness of female undergraduates. Since then it has facilitated genocide in Myanmar, the live-streaming of suicides and mass shootings, the auction of child brides in South

Sudan, foreign interference in the 2016 US presidential election, the spread of misinformation about vaccines during the Covid-19 pandemic, the attack that disrupted the peaceful transfer of power in the United States in 2021, and the work of drug cartels and human traffickers.

All of which brings us to the question of who should be thinking about technology ethics.

WHO DRAWS THE LINE – AND WHO DECIDES WHEN THAT LINE HAS BEEN CROSSED?

Technology ethics is on the geopolitical front line. What some call the 'new tech cold war' and others 'the great decoupling' between the United States and China is not only a matter of international relations, economics and trade: the data practices, tools and technologies of these two superpowers reflect their *values*.[84] Nigel Inkster, former director of operations and intelligence at MI6 (the United Kingdom's foreign intelligence service), argues that it is through technology that China's government and many of its companies are seeking to shape and reconfigure ethics norms that have previously been largely dominated by Western liberal democracies, particularly the United States.[85]

First, they are investing money and political capital in setting technology standards.[86] The information provider HIS Markit has estimated that China 'accounted for nearly half of the global facial recognition business in 2018'.[87] China is a leader not only in facial recognition and other biometric, surveillance and smart city technologies[88] but also in fintech (financial technologies), social credit scoring systems, 5G telecommunications and quantum computing.[89]

Second, China exports its technology – and thus its technology ethics – through its Belt and Road Initiative (a global infrastructure and investment project). This, as Inkster describes, has four parts: a land component (e.g. railways), a maritime component (e.g. ports), a financial component (e.g. loans) and a digital component, whereby countries all over the world use Chinese technologies to monitor, analyse and control their populations.[90] Of these, the digital component in particular has gained traction in Africa, Latin America and Europe.[91]

Third, Beijing has since 2017 been applying its ethical norms and technologies to persecute Uyghur Muslims in the Xinjiang Uyghur Autonomous Region in northwest China. The Chinese government has, for example, gathered Uyghur Muslims' biometrics (DNA, face, voice and fingerprints), forced them to download an app on their mobile phones to track their movements and monitor their social media behaviour, installed surveillance cameras in private homes, and imprisoned an estimated 1 million people in mass camps.[92] There, detainees have endured torture, rape, forced labour and the forced sterilization of women, which the United States, Canada and the Netherlands have called 'genocide' and the UK government has called 'industrial scale' abuses.[93]

How do we solve these problems – if we can even agree that they *are* problems? After all, the way that China's leaders in government and business see this framing is very different from how the leaders of Western liberal democracies see it. Other countries' leaders might see things differently too, as might any company or consumer who needs to navigate the questions of technology ethics associated with the growing US–China rivalry.

This is not just a question of institutions, statecraft and international relations. The political is also the personal. When we consider the current levels of representation in technology in universities, companies and governments along the lines of geography, sex, gender, race, ethnicity, sexuality, class, religion, language, disabilities and neurodiversity, it is evident that certain individuals and groups are shaping the creation and terms of the use of technologies much more than others are. Their dominance influences how we all think about technology ethics, whether we are conscious of this or not. It also means that some people often have a reduced voice or are excluded from conversations about how to maximize the benefits and minimize the harms of technology.

This matters when we ask who draws the line and who decides when the line has been crossed. It matters when we assess a tool or technology, because we must ask *for whom* it succeeds or fails, and with what consequences? It also matters for the global talent war. In the race for supremacy in AI, cybersecurity, quantum computing, green technology and space exploration – to name just a few of the key technologies of the future – no country or company can afford to ignore the full talent pool. Those that find ways to identify, cultivate and retain the full spectrum of human talent will have a competitive advantage over those that do not. When we think of the people who are empowered to draw the line and decide when it has been crossed, who do we see – and who is missing?

CONCLUSION

This chapter has deliberately not answered the question of where we draw the line. Instead, it has offered something more resilient: a philosophical framework that we can use to think about *how* to draw the line, *how* we (and others) can test if we have drawn it in the right place, *who* draws the line, and *who* decides when that line has been crossed. We have reflected on examples from history that demonstrate that where we draw the line can change over time. We have considered how the dominance of some countries, cultures and demographics, and the exclusion and silencing of others, is part of the problem and must be part of any solutions. Thus primed, we are ready to take a deep dive in the next two chapters into two technology ethics challenges that are shaping the lives of millions of people around the world: facial recognition technology and the new digital health tools developed in response to the Covid-19 pandemic.

PHILOSOPHY OVERVIEW

Branch: **Metaphysics.**
Key question: **What is reality?**
As applied to technology...
Questions: What is the problem we are trying to solve? What does success look like? What is the context in which we working?
Concepts: misinformation; disinformation; algorithmic decision-making; consciousness; virtual reality (VR); augmented reality (AR).

Branch: **Epistemology.**
Key question: **What does it mean to know?**
As applied to technology...
Questions: How will we know when we have solved the problem? What metrics should we (not) use?
Concepts: knowledge creation and curation; expertise; black box; transparency; explainability; accountability.

Branch: **Political philosophy.**
Key question: What is the nature of power and legitimacy?
As applied to technology...
Questions: How does (not) solving the problem affect power dynamics? How does the impact of the tool or technology change depending on who is using it, and on whom it is being used?
Concepts: how expertise is achieved, conferred and denied; technology companies as political actors; censorship versus freedom of expression; privacy; civil liberties; human rights; data ownership versus data rights.

Branch: **Logic.**
Key question: **How do we know what we know?**
As applied to technology...
Questions: How can we test if our tools and technologies are working the way we intended and/or match reality? How can we detect unintended negative consequences early on? How can we test if people are who they claim to be?
Concepts: CAPTCHA; digital identity; facial recognition and other biometrics technologies; digital health technologies.

Branch: **Aesthetics.**
Key question: **What is experience?**
As applied to technology...
Questions: How can we create a tool or technology that scales yet also protects the vulnerable and ensures accessibility and inclusivity? How do we best respond to users' needs and feedback?
Concepts: user interface (UI); user experience (UX); accessibility; value-sensitive design; personalization; data hand-over between different platforms; friction; addiction; data visualization.

Branch: **Ethics.**
Key question: **How should we live?**
As applied to technology...
Questions: How can we ensure that our tools and technologies maximize benefit and minimize harm? What values are encoded in our tools and technologies? Whose values are encoded, and whose are not? Who is investing in our technology?
Concepts: values; utilitarianism; scale; niche; panopticon; mass surveillance; AI warfare technology.

Chapter 3

Facial recognition technology

> If I am ever asked, on the streets of London, or in any other venue, public or private, to produce my ID card as evidence that I am who I say I am, when I have done nothing wrong and when I am simply ambling along and breathing God's fresh air like any other freeborn Englishman, then I will take that card out of my wallet and physically eat it in the presence of whatever emanation of the state has demanded that I produce it.
>
> If I am incapable of consuming it whole, I will masticate the card to the point of illegibility.
>
> And if that fails, or if my teeth break with the effort, I will take out my penknife and cut it up in front of the officer concerned.
>
> Boris Johnson, Member of Parliament for Henley, 2004

Nearly twenty years ago, Boris Johnson, UK Prime Minister at the time of writing, published an opinion piece in *The Telegraph* entitled 'Ask to see my ID card and I'll eat it'.[1] He was not simply opposing the idea of a national ID card, which the Labour government of the day was planning to introduce, he was stating a philosophical position – one that has a particular resonance in the United Kingdom and guides how the country answers the following three questions.[2]

- Who are you? [*Identity*]
- How can you prove it? [*Verification*]
- What can you do and not do based on your answers to the first two questions? [*Authorization*]

The answer to these questions for many countries is national ID cards, which their citizens and residents are required to use when accessing welfare services, voting or being stopped and searched by the police, among other things.

The United Kingdom has developed a different approach: we do not have a national ID card.[3]

Yes, we can apply for a driver's licence or a passport, but we only need to use them when we drive or travel abroad.

We have a National Insurance number, but we use it only to pay taxes and to access welfare.[4]

We also have a National Health Service (NHS) number, but we do not need it to book appointments with a GP or to book a Covid vaccine: we can just give our name, date of birth and the first three digits of our postcode.

We do not need ID to vote (unless we live in Northern Ireland, in which case we do). We simply register to vote, turn up to the polling station on election day, and state our name and address to the registrars, who check to see if we are on the list. If we are, we can proceed to the voting booth. This system works incredibly well: the United Kingdom has low levels of proven electoral fraud.

We do not need to show ID to the police or even tell them who we are, where we live or where we are going.[5] On the contrary, if the police want to stop and search us, *they* are legally required to tell *us* their name and police station, why they want to search us, what they expect to find and why they are legally allowed to undertake the search, and they must show us their warrant card too.[6]

Only twice in living memory has the United Kingdom deviated from its position on national ID cards. Both times occurred during a national security crisis.

The first time was in 1939, at the start of World War II, when the government passed the National Registration Act. This required every adult and child over the age of 16 to carry an ID card at all times.[7] It was an emergency measure, but it lasted well beyond the end of the war in 1945. In 1950 a man named Harry Willcox refused to show his ID to a police officer, stating: 'I am against this sort of thing.' The High Court agreed and so did Prime Minister Winston Churchill, who abolished ID cards in 1952 in order to 'set the people free'.[8] The National Registration numbers were retained, though, and repurposed as the NHS numbers that all UK residents have today.

The second time was after the attacks in the United States on 11 September 2001. The US government responded to the attacks by cracking down on civil liberties at home and by launching a global 'war on terror'. The Labour government of Prime Minister Tony Blair in the United Kingdom responded by proposing a national ID card to fight terrorism in 2003; later, it expanded this rationale to include clamping down on benefit fraud and illegal workers.[9] In 2006 the Identity Cards Bill became law, and in 2008 the first card was produced, showing the person's name, date of birth, nationality, immigration status and an electronic chip containing their biometrics, such as fingerprints and a digital facial image.[10]

Did Johnson make good on his promise to eat, masticate or cut up his ID card with a penknife? Perhaps he didn't have time to get one. Most people didn't, after all, because Labour lost the 2010 general election and the incoming Conservative–Liberal Democrat coalition scrapped the scheme and destroyed the national register with the biometric data.[11] As Theresa May, then Home Secretary and later prime minister, explained:

> This isn't just about cost savings, it's actually about the principle. It's about getting the balance right between national security and civil liberties, and that's what the new coalition government is doing.[12]

Today that balance is under threat. Biometric technologies, which turn our bodies into data and code, are making ID cards redundant. There's no need for a physical ID card when we can be identified by our DNA, our fingerprints, our faces, voices or other physical and behavioural characteristics. Of these, our face is the biometric that is increasingly used by the police and companies to identify us, often without our consent or knowledge. Yet facial recognition technology is largely unregulated, in terms of both legislation, which is inadequate, and enforcement of that legislation, which is largely non-existent.

Dame Cressida Dick, commissioner of London's Metropolitan Police (henceforth 'the Met'), acknowledges that people are starting to get worried. In 2019 she told the Lowy Institute, an Australian think tank, that:

> The next step might be predictive policing. People are starting to get worried about that ... particularly because of the potential for bias in the data or the algorithm, [such as] live facial recognition software...
>
> I'd like to talk a little bit about some of the principles that might assist with these ethical dilemmas so that we can maintain public trust and make best use of technologies, and not sleepwalk into some kind of ghastly, Orwellian, omniscient police state.[13]

She would know. Of all UK police forces, the Met has been the most enthusiastic adopter of live facial recognition technology, trialling it ten times between 2016 and 2019. In 2020 it announced that it would integrate live facial recognition technology into its operations permanently, and in 2021 it expanded this to include retrospective facial recognition as part of a £3 million, four-year deal with Japan's NEC Corporation.[14]

In doing so, the Met ignored the recommendations of the House of Commons Science and Technology Committee, which called for a moratorium on the use of facial recognition technology in 2019 and said that 'no further trials should take place until a legislative framework has been introduced and guidance on trial protocols, and an oversight and evaluation system, has been established'.[15] It ignored the independent panel that advises London's City Hall on the ethics of policing, too: in 2019 it recommended that the police only use this technology when it 'can be evidenced that using [it] will not generate gender or racial bias in policing operations' – a criterion that the Met has failed to meet.[16] It ignored the UK Equality and Human Rights Commission, which in 2020 called for facial recognition to be suspended in England and Wales.[17] And it ignored an August 2020 High Court ruling that South Wales Police's use of live facial recognition technology was 'unlawful'.[18]

Still, many of us in the United Kingdom shrug off the growing use of facial recognition technology. After all, we have lived with closed-circuit television cameras (CCTV) for decades.[19] We are already tracked 24/7 by our mobile phones and by the Big Tech companies. Millions of us put photos of ourselves and each other online. In short: it is not like we have much privacy anyway.

This chapter will test those arguments using the six main branches of philosophy that we explored in the previous chapter. We will map out how facial recognition technology challenges our approach to identity, privacy and civil liberties to see if we really are sleepwalking into some kind of 'ghastly, Orwellian, omniscient police state'. Finally, we will consider what to do about this technology, because our current approach is not fit for purpose.

METAPHYSICS: WHAT IS FACIAL RECOGNITION TECHNOLOGY?

Facial recognition technology is an example of a biometric technology: something that translates our body data from the physical world to the digital world. Biometrics include our DNA, our faces, voices, fingerprints and finger geometry (the sizes and positions of our fingers), our eyes (retinas, irises), veinprints, hands, footprints and handprints, and even our body odours and gut flora. They also include behavioural characteristics such as our emotions, personality traits and the way we write, walk and otherwise move.

In theory, *any* part of our body could be used to identify us with varying degrees of confidence – biometric technologies are not 100% accurate and are never likely to be. Our bodies are unique to us. For example, in 2012 German police solved ninety-six burglaries when they identified a burglar by his ear-prints, which he left when pressing his ears to front doors to listen for inhabitants.[20] In 2020 scientists at Stanford used anal prints to identify patients who use 'precision health' toilets.[21] Biometrics can even be used to identify individuals of other species, such as pigs, dogs and cats.[22]

Our biometrics cannot be reset, unlike a username or a password. That is a problem, because they can be stolen and gathered without our consent or knowledge.

For example, the Home Office wrongfully forced at least 449 people, including foreign parents of British children, to give DNA samples as part of their applications to live in the United Kingdom. In 2018 it apologized and investigated itself. 'I am determined to get to the bottom of how and why, in some cases, people were compelled to provide DNA in the first place,' Sajid Javid, then Home Secretary, told Parliament, discreetly omitting his department's 'hostile environment' policy towards immigrants.[23] 'Across our immigration system, no one should face a demand to supply DNA evidence and no one should have been penalised for not providing it,' Javid stated.[24]

In 2019 *The Guardian* reported that:

The fingerprints of over 1 million people, as well as facial recognition information, unencrypted usernames and passwords, and personal information of employees, were discovered on a publicly accessible database for a company used by the likes of the UK Metropolitan police, defence contractors and banks.[25]

That same year, Her Majesty's Customs and Revenue was ordered by the regulator, the Information Commissioner's Office, to delete the voice records of 5 million taxpayers because they had been collected without consent.[26]

Is it any surprise that in 2019 the Surveillance Camera Commissioner, the Biometrics Commissioner and the Information Commissioner all warned that the British government's biometrics strategy was 'not fit for purpose and needs to be done again'?[27]

Advocates of biometric technologies nonetheless argue that such abuses and errors can be prevented by better design and security practices. They have a point, but some of them also have a financial interest in making that point. As of 2019, sixty-four countries used facial recognition technology in surveillance, according to Steven Feldstein at the Carnegie Endowment for International Peace, and cities in fifty-six countries had adopted smart-city platforms.[28] The market for facial recognition technology alone will be worth $9 billion by 2022, according to estimates by Market Research Future.[29] This is part of an even bigger market for surveillance technology, which analysts at Business Research Company estimate will increase from $83 billion in global sales in 2020 to $146 billion by 2025.[30]

Not bad for something that originated in nineteenth-century France.

EPISTEMOLOGY: HOW CAN WE LEARN ABOUT FACIAL RECOGNITION TECHNOLOGY?

Facial recognition technology is based on our facial image, so let us begin there. In 1827, Joseph Nicéphore Niépce took the earliest surviving photograph: a picture of the roof at his family estate. He later collaborated with Louis-Jacques-Mandé Daguerre, whose 'daguerreotype' process made much sharper images.[31] By 1839, Alphonse Giroux had built the first commercially available photographic camera, and across the Atlantic an American named Robert Cornelius took the first known selfie.[32]

Figure 10. Robert Cornelius, self-portrait (believed to be the earliest extant American portrait photo).

Until this point, facial images were simply a record of reality. No one was using them to make decisions. That changed in 1881 when Alphonse Bertillon, a policeman and forensics pioneer, had the idea of pairing facial photographs with body identifiers such as eye colour, the shapes and angles of the ears, brows and nose (and the distances between them), and tattoos. Bertillon called his system 'anthropometry': measuring humans. Police forces around the world soon adapted the 'Bertillonage' system, including his invention of the mugshot (see figure 11).

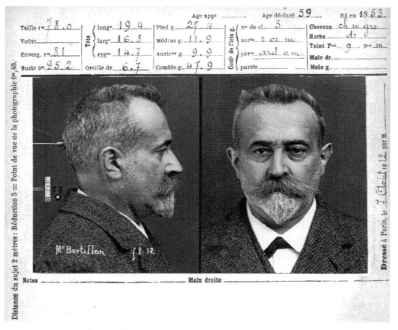

Figure 11. The anthropometric data sheet of Alphonse Bertillon.

The Paris police used the Bertillon cards to create a collection of records: a database. The French State realized that this technology had even greater potential and therefore created new forms of official identification using facial photographs: driving licences, military IDs and foreign residents' permits.[33]

Bertillon's innovative policework was even immortalized in literature in *The Hound of the Baskervilles*, when a doctor pays a visit to 221B Baker Street to ask Sherlock Holmes to help solve a mystery:

'Recognizing, as I do, that you are the second highest expert in Europe –'

'Indeed, sir! May I inquire who has the honour to be the first?' asked Holmes, with some asperity.

'To the man of precisely scientific mind the work of Monsieur Bertillon must always appeal strongly.'

'Then had you better not consult him?'[34]

Figure 12. Class studying the Bertillon method of criminal identification, circa 1910–1915.

Still, Bertillon did miss one innovation that we take for granted today: the use of fingerprints for identification and criminal investigation. That innovation came from the United Kingdom, whose colonial administrators pioneered the technique during their rule of India, and from Argentina, where a policeman in the province of Buenos Aires invented a separate system for fingerprint identification.

In the early twentieth century, countries began to include photographs alongside biometric descriptions in their passports, such as 'Forehead: broad. Nose: large. Eyes: small.'[35] By the early twenty-first century, they had added a digital component, which is why many of today's passports have a microchip containing the holder's facial biometric and, depending on the country, fingerprints and iris patterns.[36]

We are nowhere near the ethics section of our analysis yet, but already we can see that biometrics technologies are not neutral. They are tools created by police services and colonial administrators for use on criminals, suspected criminals, colonial subjects and, eventually, the rest of us.

Biometric technologies are also linked to debunked pseudo-sciences such as physiognomy – which argues that a person's character can be gleaned from their physical appearance, such as their facial expressions or simply their forehead wrinkles (metoposcopy) – and phrenology, which argues that an individual's abilities and criminality could be determined by studying the contours of their head.[37]

It is important that we face up to the origins of facial recognition and other biometrics technology because they are hard coded into many of its uses today.

For example, Guillaume-Benjamin-Amand Duchenne de Boulogne, a neurologist, published a book in 1862 that connected photography with physiognomy and phrenology.[38] As Kate Crawford notes in *Atlas of AI*, de Boulogne experimented on people that were suffering from mental illnesses and neurological conditions in the Salpetrière asylum in Paris – people who could not give true consent – and administered electric shocks to generate muscle movements and facial expressions, which he photographed and then argued were linked to emotional states.[39]

Theories linking physical appearance, emotional states and behaviour developed further in the 1880s when Cesare Lombroso – an Italian physician and professor who is known as the 'father' of criminology – argued that criminality is inherited and that 'born criminals' can be identified by their excessive tattoos, their manner of writing and talking, or the size and shape of their skulls, ears, foreheads and hands.[40] We can draw a line from this to researchers at China's Shanghai Jiao Tong University, who claimed in 2016 that they had identified features that correspond to criminality by using facial recognition technology.[41]

Plus ça change, as Bertillon might have said. While he disagreed with Lombroso's view that criminality is inherited (atavism), he did believe that convicted criminals' physical attributes could be used to predict the likelihood that they would reoffend (recidivism). This inspired him to begin taking criminals' body and head measurements and then to analyse them alongside mugshots to look for patterns.[42]

Nor was Bertillon alone in this thinking. Francis Galton – the English scientist whose research on fingerprints laid the foundations for the Galton–Henry Fingerprint System adopted by Scotland Yard in 1901 – also thought criminality could be identified from a person's features. He tried to determine the ideal 'type' of each race using composite portraits.[43] Today he is known as the inventor of eugenics: the idea of governing reproduction to improve the health and intelligence of future generations.[44] Eugenics inspired many US states to 'introduce laws for the sterilisation of criminals, the mentally ill and intellectually disabled', and it led Nazi Germany to pursue a 'racial hygiene movement, [which] led to the horrors of the extermination camps during the Holocaust'.[45]

Debunked pseudosciences should be something we only ever encounter in history, not something we find in the *New York Times* in 2021, yet in March of that year one of the founders of Clearview AI told the paper that he and Hoan Ton-That, the company's CEO, had created their facial recognition technology to explore 'physiognomy in the modern age with new technologies'.[46] This US-based company has built a database of more than 10 billion face images taken from people's social media profiles and online photo albums without their consent or knowledge. As of this writing, Clearview's database is used by more than 3,100 US law enforcement agencies and by the United States Postal Service.[47]

Not everyone is fine with this. In 2021 Canada's privacy commissioner decided that Clearview AI was a tool of mass surveillance – one that puts Canadians under a perpetual police line-up – and declared that it is illegal in Canada.[48] In Sweden, the police authority was fined by the data regulator for using Clearview's technology to identify people.[49] Regulators in the United Kingdom, Australia and France found that Clearview AI had breached their respective privacy laws and ordered the company to stop collecting images from websites and to destroy data collected in the two countries.[50] For good measure, the UK regulator also fined the company £17 million (US$22.6 million). Also that year, a number of privacy, civil liberties and human rights groups filed complaints against Clearview AI with data regulators in Austria, Italy and Greece.[51]

Yet Clearview AI is not the only company to create a search engine for faces drawn from photos of people on the internet. In 2016 Bloomberg journalist Ashlee Vance travelled to Russia, where he interviewed an unnamed developer at FindFace.[52]

> **Vance:** This idea that you're not anonymous when you walk down the street anymore, someone can just snap your photo and identify who you are. You must have thought about that when you were developing the technology.
> **Unnamed developer:** When we were developing, we were only thinking about accuracy and how cool it is. But privacy finished when you have smartphones with you, because they can track all the information about you. It's a battle between technology and privacy. My personal guess is that technology will win.

This is a weak argument: we can leave our smartphones behind, but not our faces. What is more interesting is that no one at FindFace appears to have considered if there might be any legal or ethical issues when creating a search engine for faces, or wondered who it might harm or how. The unnamed developer and his colleagues simply thought it was 'cool'.

The same applies to PimEyes, a facial recognition technology website created in Poland but recently relocated to the Seychelles. In 2021 the *Washington Post* reported that PimEyes had been used by members of the public to identify people who attacked the US Capitol on 6 January as well as to 'creep on' women and girls.[53] The company's founder told *Vice*:

> We think that face recognition technology should be available for everyone, not only for the governments, corporations, or rich people who can afford private detectives.[54]

Yet this means that PimEyes could be used not just by police but by stalkers, or by employers who wanted to look into the background of applicants and their employees, or by anyone else. The company has done nothing to prevent such uses.

The rise of companies such as Clearview AI, FindFace and PimEyes only intensifies the problems that facial recognition raises, and it poses new questions of power.

- Should *anyone* be allowed to use facial recognition technology, or only law enforcement and the security services?[55]
- What might constitute a permissible use of facial recognition technology?
- What oversight should exist?
- Is facial recognition technology simply too powerful and dangerous to be used ethically by anyone?

To help us answer these questions, let us turn our analysis to the United States, where facial recognition technology as we know it took off in the twentieth century. As Os Keyes, Nikki Stevens and Jacqueline Wernimont have shown, the training datasets created by US government agencies to develop facial recognition have at various times used images of children who have been exploited for child pornography, photos from US visa applications, pictures of people who have been arrested, and images of deceased people to test their facial recognition systems, all without the knowledge or consent of the people in those photographs.[56] Kate Crawford adds to this list images of students at Duke University who were simply walking around on campus.[57]

The attacks of 11 September 2001 led to the creation of biometric standards to track people entering and leaving the country, led by the National Institute of Standards and Technology (NIST), a physical sciences laboratory that is part of the US Department of Commerce.[58] A technology created for use by the military, the intelligence services and law enforcement was therefore extended to immigration and border control under the auspices of the 'war on terror'. Although the United States has now largely withdrawn from Iraq and Afghanistan, components of the war on terror infrastructure, such as facial recognition technology, remain. They are also expanding.

China – whose use of biometric and surveillance technologies was described in the previous chapter – has made real the nightmare scenario that Brad Smith, Microsoft's president, outlined in 2019:

> A government could use facial recognition technology to enable continuous surveillance of specific individuals... [It] could follow anyone anywhere, or for that matter, everyone everywhere. It could do this at any time or even all the time. It could unleash mass surveillance on an unprecedented scale.[59]

To which we can only reply: 'Then why is Microsoft building it?!' No one is forcing them; they are choosing to build it.

Facial recognition is therefore not just a biometrics technology: it can also be a technology of surveillance and control. It is not simply a tool that can be used for good or bad, depending on who uses it. It is complex, and its myriad uses have effects ranging from the harmless to being what Cathy O'Neil calls a 'weapon of math destruction': an algorithmic technology that can harm at mass scale.[60] Only by distinguishing between those effects can we target a response to minimize the technology's harms.

LOGIC: HOW DO WE KNOW WHAT WE KNOW ABOUT FACIAL RECOGNITION?

When does a tool or a technology become a weapon – particularly a 'weapon of math destruction'?

Sometimes we regulate a tool or technology, such as certain knives, all guns, cars, nerve agents and atomic bombs. At other times we prefer to regulate *the ways in which certain tools and technologies are used*.[61] For example, we do not regulate forks (a tool) but we do regulate using a fork to stab someone to death (a way in which a tool or technology is used). Precision matters when defining the problem we are trying to solve.

In this section I will analyse some of the most common uses of facial recognition technology – though by no means all of them – and assess the risk of harm that each of them poses. This will allow us to see whether a moratorium or a ban is warranted, or if it is possible to devise a more targeted response to mitigate harm. Table 1 shows an overview of which uses I think are most in need of regulation.

1:1 face matching

We use 1:1 face matching any time we match our face against a facial image stored either on a device we own, such as a smartphone, or on a centralized database run by someone else, such as our government.

Table 1. Uses of facial recognition technology and their risk of harm.

Type of facial recognition technology	Example of how it is used	Risk of harm
1:1	• To unlock our smartphone	Low
	• To access government services (e.g. Aadhaar)	**High**
	• To pay with our face	*Medium* (**high** for children)
	• To monitor workers	**High**
	• To enter a building	**High**
	• To access humanitarian assistance	**High**
1:many	• To tag/be tagged on social media	*Medium*
	• To identify a person in a crowd	**High**
	• As a tool of foreign policy	**High**
	• As a tool with which to wage war	**High**
Dual use 1:1 and 1:many	• To control borders and immigration	Low
Face analysis: physical	• To apply online for a passport	**High**
	• To analyse our physical health	*Medium*
Face analysis: classification	• By ethnicity and race	**High**
	• By sexual orientation or political orientation	**High**
	• By emotional state	**High**

To unlock our smartphone

To set this up, we must first capture our raw facial biometric using our smartphone's camera. The software in our smartphone then takes our raw facial biometric, turns it into a map of many points and then translates it into a 'hash': a mathematical representation of our face. In other words, our smartphone turns our face into code. This hash never leaves our device.

With facial verification technology, our smartphone uses 1:1 mapping to verify that *this* face is allowed to unlock *this* device.

It does this by comparing our face in real time against the hash of our face, which is stored only on our phone. If they match, the phone unlocks. If they do not match, the phone stays locked.

In most cases, this is done with our knowledge and consent, because we are the ones doing it. However, we should acknowledge the possibility that someone could force us to use our faces to unlock our phones, just as they could force us to enter the passcode. Nothing on a phone is entirely secure.[62]

Risk level	Reasoning
Low	• Here 1:1 matching is done with our knowledge and consent • In the overwhelming majority of cases we only use this technology on ourselves and on our own device; however, it is possible use this technology on other people

To access government services
Several countries use facial verification technology as the means of enabling citizens to access government services. For example, India's government has built 'the world's largest biometrics project', a multimodal biometrics solution (iris, fingerprints, face) of more than 1 billion Indian citizens and residents' data known as Aadhaar.[63]

This is an example of 1:1 face matching: once a person has had a record created of them, all they have to do to access government services is allow the Aadhaar system to check their face (or iris or fingerprint) against the original record in the database.

Aadhaar solves some real problems, as many people in India do not have any official identification, but it causes problems too.

For example, it sets the default to be opting in, rather than opting out, and it offers little to no choice or informed consent to some of the country's most vulnerable people: those who have no other form of identification. The Indian government made signing up to Aadhaar voluntary, and compelling Indians to use Aadhaar became illegal in 2017 after the country's Supreme Court ruled that the 'right to privacy is an intrinsic part of right to life'.[64] However, it is effectively mandatory for anyone who needs to access government services, open a bank account or obtain a mobile phone contract.[65]

It is not secure. In January 2018 reporters at *The Tribune* newspaper in India paid 500 rupees (a little less than $8) to obtain a login and password that granted them access to the names, addresses, postal codes, photos, phone numbers and emails of every person in the system. For just 300 rupees more, the reporters could print out copies of anyone's unique identity cards – and start using them.[66] In May 2018 Bloomberg reported another data leak of citizens' unique twelve-digit Aadhaar numbers, bank details and identity and demographic details, including their photographs.[67]

It endangers people's physical safety. India's state governments have been using citizens' Aadhaar numbers as a unique identifier while tracking their caste, religion and geolocation, and they have published this information on a public online dashboard.[68] In a country with a history of intercommunal violence leading to deaths and destruction, this makes it easier than ever to search for people.

It has normalized facial recognition technology in a country that, as Madhumita Murgia notes, has no national law to define limits on its use. [69] In 2021 Indian Railways deployed nearly 500 facial recognition cameras (made by the Russian start-up NtechLab) to track millions of commuters each day, and the government is seeking bids for an integrated National Automated Facial Recognition System.

Risk level	Reasoning
High	• This type of 1:1 matching is done with our knowledge
	• In theory it offers the possibility of consent but in practice it is often mandatory
	• There have been problems with the security of the system, although that is also true of many government ID systems
	• India's use of this system to track people's caste, religion and geolocation puts them at physical risk, and shows the dangers associated with what data is included in any national biometric scheme

To monitor workers

In 2019 fourteen people who worked for Uber Eats told WIRED that they had been wrongfully suspended or fired because they had failed the company's 'Real Time ID Check', which uses 1:1 facial verification technology to check that its drivers are not subcontracting their shifts to people who have not passed background checks or are otherwise not entitled to work.[70]

Here is how the system works. When drivers sign up to Uber, their face biometrics are stored in a database. When they try to log in to their shifts, they are asked to take a picture of themselves – a 'selfie'. Uber then checks to see if the driver's selfie matches the facial biometric stored on their profile in Uber's system. In the case of these fourteen drivers, Uber's system was too aggressive and resulted in a false negative (denying access to a person who should have been authorized) rather than a positive, which is what it should have returned. The question is: why?

The couriers, who were all from Black, Asian and minority ethnic groups, said that the software had been unable to recognize their faces. Researchers have long established that facial recognition technology does not work as well on people with darker skin, and performs worst of all on women with darker skin.[71] As Professor Peter Fussey, a sociologist at the University of Essex who specializes in facial recognition technology, told WIRED:

> There is no facial recognition software that performs equally across different ethnicities. That technology doesn't exist. If you're bringing in that technology into an already unequal environment it just exacerbates those conditions and amplifies racial inequality.[72]

This poses a risk to some employees more than others, and is thus an issue for employment law as well as for unions. It is also part of a growing trend that affects many workers: surveillance at work. Again, this is a question of power: it is unlikely that any executive in any organization would ever agree to the surveillance that they routinely impose on workers lower down the hierarchy.

For example, Amazon requires its delivery workers to submit a selfie via an app before each shift and it also monitors their driving, phone use and location.[73] In 2021 the company installed a four-lens, AI-powered camera in its US delivery vehicles to record and monitor drivers' faces and bodies during their shifts. Employees had a choice: 'consent' or find another job. That is not consent; that is an ultimatum. Amazon also reserved the right to share this information about their workers with third-party service providers and Amazon group affiliates.[74] It has turned its workers into a commodity whose data can be shared, used and monetized. The only way they can refuse is by quitting.

Finally, during the Covid-19 pandemic, when many office workers worked from home, some employers increased their existing surveillance practices – keystroke tracking, screenshots, video and audio recording – to include face-scanning software to verify their employees' identities and monitor their productivity.[75]

Risk level	Reasoning
High	• This type of 1:1 matching is done with our knowledge but not with meaningful consent, as workers may feel pressured to agree in order to keep their job
	• Workers with darker skin are particularly at risk from facial verification software
	• The use of biometrics technologies, even when supposedly for health and safety purposes, may violate workers' rights, as may the sharing of workers' data with third parties and affiliates

To pay for things
Facial verification technology can be used to pay for things when we are out and about (as opposed to when we pay for things

online).[76] It is probably safe for patrons of Daddy's Chicken Shack in southern California to enjoy the convenience of 'pay with your face' technology because it is a novelty in most of the United States, as it is not required in that country. Yet.[77]

By contrast, in China, as the *Financial Times* has reported:

> Face scans have replaced or augmented human identity checks in hotels, boarding flights and trains, and at banks and hospitals. Regulations require telecom carriers to scan the faces of users registering for mobile phone services.[78]

This is not so much 'pay with your face' as 'pay with your face to conduct even the most basic transactions because you have no choice and will be penalized if you refuse' – which is admittedly less catchy.

The risk of this technology is especially high for children. In 2021 the *Financial Times* reported that a number of British schools were scanning the faces of thousands of children to pay for lunch in their canteens.[79] Following a backlash from privacy and children's advocates and intervention by the Information Commissioner's Office (ICO) and the Biometric and Surveillance Camera Commissioner, some of the schools halted their use of the technology – temporarily. There is nothing to stop them from resuming, should they wish. For this reason, Lord Clement-Jones secured an emergency debate in the House of Lords, during which Lord Scriven[80] proposed that Parliament pass new legislation specifically to protect children:[81]

> It does not have to be like this. We can step back from allowing technology to lead the debate. We can step back from children being normalized into their bodies being used to access school services, and we can move forward with asking where we, as a country, draw the line, and bring forward legislation to show that there is a line. I suggest that the

line is the use of biometric technology in schools on young people.

Risk level	Reasoning
Medium (**high** for children)	• This type of 1:1 matching is done with our knowledge
	• Depending on the country and the organization, it can be done with or without our consent
	• Assuming that the people running the database are using good security practices, our facial image should be stored as a hash and not as a raw biometric. The hash itself is not a security risk, because if it is stolen, it is not possible to reverse engineer the original face from it or to reuse it
	• However, if the people running this system also store the original image(s) and not just the hash, that *would* be a security risk as it could be stolen and/or reused; of course, we have no way of checking how our data is being handled and whether only a hash is stored
	• Pay with your face is fine as long as people always have other options that are just as easy to use and face no penalty for using those instead

To enter a building

Imagine that *someone else* wants to use our face to decide if we are allowed to do something, such as enter a building. They would have to compare our face in real time against a database of facial images that may be stored on a privately owned server or else in the cloud. If our face matches one of them, we are allowed to enter the building, but if it does not, we are refused entry.

Nelson Management Group, the owners of the Atlantic Plaza Towers in Brooklyn, proposed using this form of facial verification technology in 2019 as a security measure to control who was allowed in and out of the building complex.[82] Rather than use a key or a fob to unlock the door to the building, tenants would have been required to allow a camera to take a photo of their face, which would then be compared against a database of facial images. This did not go down well. The residents protested, the landlord reversed its decision, and State Assemblywoman Latrice Walker and Congresswoman Yvette Clarke both introduced No Biometrics Barriers to Housing acts, which as of this writing are working their way through the legislative process.[83]

Here, facial verification technology was not simply about determining access to a building: it was about power, and specifically the balance of power between a landlord and their tenants. This requires us to acknowledge class and wealth (who owns property and who does not). It also requires us to acknowledge race: as Mutale Nkonde notes, 90% of the Atlantic complex's residents are people of colour.[84] She reminds us that racism is not just about the present day; it is rooted in our histories and cultures:

> There is historical precedent for technology being used to survey the movements of the Black population. In 1713, New York passed the Lantern Law which demanded that any enslaved person over the age of 14 carry a lantern at night so that they could easily be seen by White people.

As Simone Browne, author of *Dark Matters: On the Surveillance of Blackness*, explained:

> Any white person was deputized to stop those who walked without the lit candle after dark. So you can see the legal framework for stop-and-frisk policing practices was established long before our contemporary era.[85]

The argument for banning this use of facial verification technology is strong because it relates to social housing. By contrast, employers who wish to use it could argue that they already require employees to show ID cards to enter and leave a building – in other words, they already surveil their workers' movements. Yet workers could counterargue that this technology is unnecessary and disproportionate, given that many other non-biometric options (e.g. a badge or a lanyard) work perfectly well.

Risk level	Reasoning
High	• 1:1 matching reflects an imbalance of power, in which one person or organization can force others to use their face to transact; they have knowledge that their face is being used but they have not given consent • Even if 1:many matching (described in the next section) is done with our knowledge, we may not have the power to withhold our consent, which risks harm to us (though not to our landlord or employer)

To access humanitarian assistance
The United Nations Refugee Agency has been using a Biometric Identity Management system to collect the face and fingerprint biometrics of refugees since 2013. In the programme's first five years, it acquired the data of 4.4 million adults and children over the age of five out of a total global refugee population of 22.5 million.[86]

By definition, refugees are vulnerable: they are fleeing war or other crises; they are in desperate need of food, shelter, clothing and medical care; and they are unable to appeal to elected representatives to advocate on their behalf. Sometimes they lack identity documents, have become separated from family members, or have died and need to be identified.

Ben Hayes and Massimo Marelli note that this population is unable to give true consent. At the same time, humanitarian organizations are under pressure to use biometrics to distribute aid efficiently, fight fraud and corruption, and help governments and international organizations such as the United Nations to control migration and fight terrorism.[87] This creates a risk that refugees' biometric data will be taken without their consent and shared with governments and other international organizations.

In recognition of these risks, in 2015 Oxfam imposed a voluntary moratorium on the use of biometrics in its work to combat poverty, and in 2021 it released the self-imposed guidelines under which it will collect biometrics from vulnerable people.[88] By contrast, the UN Refugee Agency continues to collect refugees' biometrics. So does the International Committee of the Red Cross (ICRC), but only for forensics and the restoration of family links, not for the distribution of aid.[89]

Hayes and Marelli explain how the ICRC developed a Biometric Policy in 2019 to take into account these various concerns, pressures and ethical questions.[90] It then created a 1:1 face matching system in which a person's face would be matched against the biometric template stored on a smart card. This allowed the adoption of a token-based system in which biometric data was stored on a card but *not* in a centralized biometric database. Cardholders could withdraw or delete their data at any time and they could return or destroy the card – they could also refuse to provide their biometric data in the first place.[91]

Risk level	Reasoning
High	• This type of 1:1 matching is done with people's knowledge but may not always be done with their consent, as they are extremely vulnerable and have little power and few rights to protest against the collection and analysis of their biometrics
	• The use of biometrics in humanitarian assistance requires urgent attention, and likely new legislation, from elected officials in liberal democracies in consultation with humanitarian organizations, researchers and other stakeholders

1:many face matching

1:many matching compares images of our face against those stored in a database or in the cloud. We may know about this, and we may have consented to it, but it can also be done without our knowledge or consent.

To tag someone on social media

Until recently it was not uncommon to log into our social media account and discover that someone had tagged us in a photograph, which would then pop up in our notifications as well as those of people in our network. How we felt about this depended on how we felt both about the person tagging us and about the image. Our only option if we wanted to un-tag ourselves or have the image taken down was to contact the person who had tagged us in the first place.

Recently, social media companies have added functionality to give us a bit more control. For instance, we can adjust our settings so that we can never be tagged in an image, or so that our friends must first obtain our permission. We can request that photos be removed or we can report them so that they are taken down.

What inspired this change? 'It's complicated,' as Facebook likes to say. It might have something to do with a class action lawsuit filed in 2015 by 1.6 million Facebook users based in Illinois, the US state with the strongest biometric protections: the 2008 Biometric Information Privacy Act.

In 2021 a US federal judge approved a $650 million settlement against Facebook for failing to get consent from those users before using facial recognition technology on their photos.[92] The judge's ruling underscores the power of facial recognition technology and explains why it is not like CCTV:

> The facial-recognition technology at issue here can obtain information that is 'detailed, encyclopaedic, and effortlessly compiled,' which would be almost impossible without such technology...
>
> Once a face template of an individual is created, Facebook can use it to identify that individual in any of the other hundreds of millions of photos uploaded to Facebook each day, as well as determine when the individual was present at a specific location.

Facebook can also identify the individual's Facebook friends or acquaintances who are present in the photo...

[It] seems likely that a face-mapped individual could be identified from a surveillance photo taken on the streets or in an office building. Or a biometric face template could be used to unlock the face recognition lock on that individual's cell phone.[93]

In November 2021 Facebook announced that it was ending its use of facial recognition and deleting the facial data of more than a billion people (it will still use the technology to help people gain access to a locked account, verify their identity in financial products or unlock a personal device).[94]

Nor is Facebook the only tech giant against which the Biometric Information Privacy Act has been used to great effect. As Woodrow Hartzog notes, Clearview AI cited the Act in May 2020 when it announced it was ending all service contracts with all non-law-enforcement entities based in Illinois. The only reason Clearview AI did this was to try to avoid an injunction and potentially large damages.[95] Alas, their strategy may yet fail: as of this writing, the American Civil Liberties Union is suing the company.[96]

Risk level	Reasoning
Medium	• Not all Facebook users live in Illinois, so unless the other forty-nine US states enact stronger biometrics protections – to say nothing about the hundreds of millions of Facebook users around the world – the company (and others that use facial recognition technology) remains free to harvest people's facial images

To identify a person in a crowd
Facial recognition technology is used to fight crime, such as when it spots shoplifters,[97] to find missing persons and to identify suspects and criminals. It has also been used during sports

and music events to search for known troublemakers, suspected criminals and celebrities.[98]

Many people will be fine with these uses. The risk depends on what activities are criminalized.

For example, in Russia the authorities have been using Moscow's facial recognition camera network to identify and detain people who attend protests in support of Alexey Navalny, who President Vladimir Putin has tried to assassinate and currently has imprisoned.[99] According to Bloomberg, some of the people who have been detained are journalists who were attending the protests in a professional capacity, and the authorities are also investigating lawyers and doctors who provided professional assistance to opposition activists.[100] Most of the protesters were tracked back to their homes – *after* the protest.

Criminalizing protests and journalism creates a chilling effect on freedom of speech and assembly, as Mikhail Biryukov, a lawyer who represents several of the detained, told Bloomberg: 'They use various techniques but they have one goal: intimidation. Creating uncertainty over when they might come for you can be a better deterrent than using violence.'[101]

Yet facial recognition technology does not only chill dissent and protest in countries, such as Russia or China, that more commonly use authoritarian mechanisms. It can have this effect in liberal democracies too.[102]

As Clare Garvie, a senior associate at Georgetown University's Center on Privacy and Technology, explains, facial recognition technology is 'a forensic without a science'.[103] This refers to the risks both of misidentification and of wrongful conviction, as well as to the rights of the accused to information about how they were identified. According to Garvie, 'Face recognition has been used in thousands of cases across the United States, and in most of these cases the accused never had the opportunity to challenge it.' She argues that this amounts to a violation of the right to a fair trial.[104]

Risk level	Reasoning
High	• This type of 1:many matching is done without our knowledge and consent
	• Anyone can use these tools, not just law enforcement; this raises the risk of harassment online and in the real world, including stalking or the use by employers and potential employers to research employees' and applicants' backgrounds
	• Not all of these tools have been submitted to the US National Institute of Standards and Technology to test their accuracy

As a tool of foreign policy

According to Steven Feldstein of the Carnegie Endowment for International Peace, in 2019 at least 75 out of 176 countries were 'actively using AI technologies for surveillance purposes', including smart city/safe city platforms (56 countries), facial recognition systems (64 countries) and smart policing (52 countries).[105]

China dominates this market and uses it as a foreign policy tool through its Belt and Road Initiative, which involves infrastructure projects and loans.[106] Chinese companies such as Huawei, Hikvision, Dahua and ZTE – all of which are considered national security threats by the US government – supply AI surveillance technology in sixty-three countries, thirty-six of which have signed onto the Belt and Road Initiative.

In 2019 the *Financial Times* reported on a deal between CloudWalk, a Chinese facial recognition company, and the government of Zimbabwe to share data on millions of African faces to train CloudWalk's algorithms.[107] CloudWalk hopes to do better in solving the problem of misidentification and mislabelling of people with darker skin that undermines a lot of facial recognition. Even Facebook and Google, which have some of the largest available datasets of images on which to train their algorithms, have had to apologize after their AI put a 'primates' label on a video of black men interacting with white civilians and police officers (Facebook, in 2021) and labelled pictures of black people as 'gorillas' (Google Photos, in 2015).[108]

Huawei provides AI surveillance technology to at least fifty countries worldwide – 'No other company comes close,' Feldstein notes.[109] The next largest non-Chinese supplier of AI surveillance technology is Japan's NEC Corporation, which provides its technology to fourteen countries, including the United Kingdom, and specifically the London Metropolitan Police.[110]

Risk level	Reasoning
High	• This type of 1:many matching is done without people's knowledge and consent
	• In liberal democracies, people have rights to privacy and civil liberties; however, these rights are under threat from both the police and private companies, and they are not always enforced, despite these systems being subjected to oversight from elected representatives, civil servants and regulators
	• When liberal democracies use technology provided by companies who are associated with human rights abuses and which are also compelled to share data with their home government, they legitimize these activities

As a tool with which to wage war
Facial recognition technology is a dual-use technology, which means it can be used for either civilian or military purposes.[111]

NATO forces have used facial recognition technology in Iraq and Afghanistan since the attacks of 11 September 2001.[112] They have collected photographs, facial scans, eye scans, fingerprints and DNA from millions of people because these biometrics offer an unparalleled advantage over other forms of identification. As Sergeant Major Robert Haemmerle explained to the *New York Times*: 'You can present a fake identification card. You can shave your beard off. But you can't shave your biometrics.'[113]

People under military occupation are extremely vulnerable. They have few rights in general, and they have even fewer when it comes to refusing to have their biometrics taken. Nor do they have any say over how their data is used or with whom it is shared during a conflict or after one. As the US withdrew from Afghanistan in August 2021, Human Rights First warned that the Taliban had probably gained access to biometric databases and equipment, raising the risk that they could use the face, fingerprint and iris scans from those databases to track down Afghans who had opposed them during the twenty-year NATO occupation.[114] As of this writing, neither US nor NATO officials have explained what will be done with the biometric data they collected, or what steps they have taken to ensure it cannot be used to harm the Afghan people.

Risk level	Reasoning
High	• This type of 1:many matching can be done with people's knowledge but without their consent (e.g. if they are asked/ordered to submit to this by military occupiers) • There are no rules governing how this data can be used or the conditions under which it must be destroyed, which means it could be used to train AI for military and other purposes

Dual-use 1:1 and 1:many face matching

To control borders and immigration
After the attacks of 11 September 2001, the US Congress authorized the collection of biometrics, including facial images, from non-citizens only.[115] As a result, US Customs and Border Control has since 2004 been allowed to collect, use and retain for seventy-five years – so essentially, for life – the facial images of non-US citizens who arrive in the United States.[116]

As of 2019, the United States was using facial recognition technology in seventeen US international airports, with three more 'in the works', and it aims to use it on 97% of departing passengers by 2023 to search for known troublemakers and suspected criminals (1:many matching) and to speed up the time it takes travellers to get from curb to gate (1:1 matching).[117]

Furthermore, in 2021 newly elected US President Joe Biden introduced the US Citizenship Act of 2021 to, in his words, restore 'humanity and American values' to the US immigration system. Who knew that humanity and American values include 'cameras, sensors, large-scale X-Ray machines and fixed towers'? And as Niamh Kinchin argues, this new system probably also includes infrared cameras, motion sensors, facial recognition, biometric data, aerial drones and radar.[118]

Still, the United States is far from alone in using biometrics and surveillance technologies to control immigration. Accenture* has an entire business dedicated to border service technologies and it has helped to build

- the US Visitor and Immigrant Status Indicator Technology,[119] the biometric identity ePassport scheme at London's Heathrow airport,[120] and the European Union's Biometric Matching System and Visa Information System;[121]
- the Biometric Identity System for the United Nations High Commissioner for Refugees, which provides a digital identity (face scan and fingerprints) for 33.9 million refugees across 125 countries;[122]
- a 'Video Analytics Service Platform in Singapore to connect to existing and new sensor infrastructures (including dozens of CCTV cameras), apply computer vision and predictive analytics to surveillance video feeds to detect various events, and generate business alerts for six different government agencies';[123] and
- the United Kingdom's 'Brexit app',[124] which processes the biometric and other highly sensitive personal data of EU nationals wishing to remain after the country's exit from the European Union.

---

* Disclosure: the author worked at Accenture from 2000 to 2003 and at Accenture Research from 2017 to 2018, albeit not on any of these projects.

That is just one company, and only some of that company's activities in 'border technology' and 'digital identity'. There are many more. Such companies claim to be selling products and services, and they are. But in doing so, they are also building an invisible infrastructure that increasingly monitors and controls our movements, with little or no transparency and accountability.

Risk level	Reasoning
Low	• This type of 1:many matching is done with our knowledge
	• It is not done with our consent, though, as biometrics technologies are a requirement to fly in many countries and with a number of airlines
	• However, the biometrics technologies used in these activities are subjected to testing and oversight
	• In liberal democracies, these systems are further subjected to oversight from elected representatives, civil servants and regulators

Face analysis: physical

Facial recognition technology can do more than simply identify and verify us: it can also be used to analyse our physical health and our characteristics in order to classify us according to certain criteria. It allows others to make decisions about us – sometimes with our knowledge and consent, often without.

To apply for a passport
In October 2020 the BBC investigated the Home Office's online face checker, which checks only if a person has submitted a valid passport photograph (e.g. eyes open, mouth shut, no smiling). It found that it did not work as well on people with darker skin, with the result that their passport applications were more likely to be rejected.[125]

To find out why this was happening, the BBC submitted 1,000 photographs of politicians from around the world to the Home Office's system and reported the following results.

- 'Dark-skinned women were told their photos were poor quality 22% of the time, while the figure for light-skinned women was 14%.'
- 'Dark-skinned men were told their photos were poor quality 15% of the time, while the figure for light-skinned men was 9%.'
- 'Photos of women with the darkest skin were four times more likely to be graded poor quality than women with the lightest skin.'

One black student, Elaine Owusu, was told that her photo was rejected because it looked as though her mouth was open – even though it was closed. Joshua Bada had the same problem when he applied for his passport using a high-quality photo booth image: the Home Office system could not read his facial image correctly, thought his closed mouth was open, and rejected his application.[126]

The Home Office had known about this problem since 2019, when it ran trials, but it decided to launch the system anyway.[127] The next year, in response to Freedom of Information requests, the Home Office gave the following explanation:

User research was carried out with a wide range of ethnic groups and did identify that people with very light or very dark skin found it difficult to provide an acceptable passport photograph. However, the overall performance was judged sufficient to deploy.

This is a powerful example of how utilitarianism's aim to maximize benefits and minimize harms is flawed. The Home Office is trying to deliver passport services as quickly and cost effectively as possible, which is commendable. However, in deploying facial verification technology that it knew discriminated against people with darker skin, it transferred the cost of its 'improvement' onto a minority of citizens, knowing that they would have to go through a confusing, frustrating and even painful experience of technological and bureaucratic racism. It could have waited to deploy facial verification technology until it worked to a high degree of accuracy for all UK citizens, not just those that fell within a range of skin tones. It chose not to. That says nothing about the technology – it says everything about the government's values.

Risk level	Reasoning
High	• Here, facial analysis is done with our knowledge and consent (at the moment there are non-technological options to apply for a passport) • This technology does not work as well on people with darker skin, subjecting them to an inferior experience of government than other citizens

To analyse our physical health
Facial analysis can be used to monitor blood pressure, cardiovascular disease, dementia, type 2 diabetes, or to check for drowsiness or inebriation in drivers, pilots or anyone operating machinery.[128]

Risk level	Reasoning
Medium	• If we know about this and have given our consent, it could be OK • However, there is a risk that workers could be pressured into agreeing to this and not have the right to refuse without penalty • If this is used on us without our knowledge and consent, the potential for abuse is high; this constitutes health data, which is protected information; in countries without universal health coverage, such as the United States, it could be used to deny someone healthcare if it were to detect a pre-existing health problem

Face analysis: classification

Classification of people based on characteristics may seem straightforward until we remember that ethnicity and race are constructed and highly contested categories.[129] How we identify ourselves may not be the same as how we are identified by facial technologies.[130] This also applies to other ways people might choose to classify us, from our sexual orientation to our political orientation or even our emotions. We will likely never know how we are being classified, and for what purpose.

By ethnicity and race
Chinese technology companies such as Huawei, Dahua, Uniview, Megvii, Alibaba and Hikvision have all developed technology that claims to be able to identify if someone is a Uyghur, the ethnic group that the Chinese government has been persecuting for years, above and beyond the surveillance that they already use on their general population.[131] Beijing has ordered these companies to draw up standards for facial recognition systems that specify how data should be segmented by characteristics such as ethnicity.

IBM developed something along similar lines for New York City in 2018 when it created software that allowed the police to search surveillance video footage based on skin colour.[132] This followed its role from 2012 to 2016 in building a video surveillance and police analytics system in Davao City in the Philippines, a country where President Rodrigo Duterte has deployed death squads to murder 'street children, drug dealers, drug users, petty criminals, anti-Duterte activists, elected officials and Catholic priests'.[133]

In 2020, after the killing of George Floyd by the police officer Derek Chauvin in the United States, IBM announced that it was getting out of the facial recognition technology business.[134] However, it is unclear exactly what this means. The technology it has already built for others still exists. It may have contracts that it needs to fulfil and support. All we can know from their announcement is that the company will not continue to sell facial recognition technology in the future – unless it changes its mind, of course.

Risk level	Reasoning
High	• This type of face classification is done without our knowledge and consent
	• We have no way of vetting or challenging any claims, including whether we agree with how we are classified
	• While this technology could be used to fight crime and terrorism, it can also be abused by law enforcement, as the examples above show, including in liberal democracies

By sexual orientation or political affiliation
As Kate Crawford has observed: 'The great majority of university-based AI research is done without any ethical review process.'[135]

Certainly it boggles the mind how two researchers at Stanford University got permission in 2017 to find out if they could use AI to identify people's sexual orientation based on their faces.[136] There is a problem in even asking the question: it legitimates physiognomy as a form of scientific inquiry and imposes a contingent, socially constructed division of sexual attraction onto heterosexuality and homosexuality. That is dangerous in a world in which sixty-nine countries still criminalize homosexuality.[137]

'The ethical implications of developing such systems for queer communities are far-reaching, with the potential of causing serious harms to affected individuals,' Google's DeepMind researchers warned in a paper in 2021 that considered the effects of AI on people who identify as lesbian, gay, bisexual, transgender or asexual.[138] 'Prediction algorithms could be deployed at scale by malicious actors, particularly in nations where homosexuality and gender non-conformity are punishable offenses.'

In 2021, Michal Kosinski, one of the Stanford University researchers who sought to find out if our sexual orientation can be determined from our face, also explored whether our faces reveal political orientation.[139] He argues that they can, and that the privacy threats posed by facial recognition technology are 'unprecedented'.

Again, Kosinski's research is problematic because of what it claims to find but also simply for the fact that it is being done at all. People in many parts of the world are persecuted for their political beliefs. Any suggestion that political beliefs could be predicted on the basis of someone's facial features risks giving further ammunition to governments and political regimes to cause harm.

Risk level	Reasoning
High	• This type of face classification can be done without our knowledge and consent, and we have no way of vetting or challenging any claims
	• Even if it is done with our consent, it is highly questionable as a premise (e.g. can it do what it claims to do?) and even more so on ethics grounds (e.g. what harms could result from this research or use, and in what scenario(s) would it ever be ethical to do this?)
	• Given the number of countries around the world in which homosexuality is criminalized, this technology – whether it works or not – could be used to persecute people both in countries where homosexuality is criminalized and where it is legal but homophobia exists

By emotional state

Emotion detection technology, also known as affective technology, is among the most controversial face technologies.[140] Many researchers have called for it to be banned, yet it is used by companies to vet job applicants; by airport security to assess passengers; by police to judge the mood of a crowd; and by schools to determine whether children are learning and, in the United States, to judge how likely they are to open fire on their classmates and teachers.

Emotion detection technology is based on two assumptions. First, that all humans have a set of universal emotions. Second, that it is possible to read our emotions from our faces. However, it is far from clear that humans have a set of universal emotions. On the contrary, there are diversity and complexity in humans' emotional ranges, as well as individual and cultural differences in how we express emotions. As for the claim that it is possible to predict our emotions from reading our faces, Crawford notes that:

A comprehensive review of the available scientific literature on inferring emotions from facial movements published in 2019 was definitive: there is *no reliable evidence* that you can accurately predict someone's emotional state from their face.[141]

None of this stopped HireVue, a recruiting-technology firm, from using emotion detection technology on behalf of some of the biggest companies in the United States, including Hilton, Unilever, GE and Delta Airlines.[142] As the *Washington Post* reported, job candidates were not informed of their 'employability score' or what they did 'wrong', nor could they find out what they could have done differently to get a better score.

This led the Electronic Privacy Information Center (EPIC) to file a complaint in 2019 with the US Federal Trade Commission to investigate HireVue for 'unfair and deceptive' practices. HireVue's system's 'biased, unprovable and not replicable' results, EPIC officials argued, were a major threat to workers' privacy and livelihoods. In January 2021 HireVue said that it would stop its practice of using facial analysis on people in video job interviews.[143]

Risk level	Reasoning
High	• This type of face classification is done without our knowledge or consent • It only recognises a limited number of emotions, and even these it may misidentify • It is based on dubious and debunked theories and research and ignores the many reasons why people show emotions differently

We can use the overview table on page 101 to support a conversation about what we want to do about the different uses of facial recognition technology and their associated risks, which brings us neatly to how facial recognition technology looks through the lens of political philosophy and the exercise of power.

POLITICAL PHILOSOPHY: HOW DOES FACIAL RECOGNITION TECHNOLOGY AFFECT POWER DYNAMICS?

Countries around the world are responding very differently to facial recognition technology and to how it has changed the balance of power in their societies. Here we focus on the differences between the way this technology is used and regulated in the United States and the United Kingdom.

United States
In the United States, researchers Joy Buolamwini, Deborah Raji and Timnit Gebru have investigated the facial recognition systems built by Amazon, IBM and Microsoft. 'The companies I evaluated had error rates of no more than 1% for lighter-skinned men,' Buolamwini wrote in *TIME* magazine. 'For darker-skinned women, the errors soared to 35%.'[144]

The US National Institute of Standards and Technology, the federal laboratory that develops standards for new technology, also found 'empirical evidence' in 2019 that most of the available facial recognition algorithms exhibit 'demographic differentials' that can worsen their accuracy based on a person's age, gender or race.[145]

Giving the police a technology that routinely misidentifies people is a recipe for trouble in any country, but this is *especially* true in the United States, where in 2015 police killed more people in a matter of days than other countries do in years.[146] Police use of facial recognition technology has already led to the wrongful arrest of at least three black men as well as the misidentification of a US-born Muslim student at Brown University in Rhode Island who was mistakenly accused of being a Sri Lankan bombing suspect.[147] No wonder Axon, the largest provider of police body cameras in the United States, decided in 2019 not to equip its cameras with facial recognition, citing ethical concerns.[148]

Microsoft and IBM responded to the researchers' criticisms by upgrading their systems, but Amazon defended its Rekognition software and continued to sell it to US law enforcement even after it misidentified twenty-eight members of the US Congress – mainly Black and Latinx – in a test conducted by the American Civil Liberties Union in 2018.[149]

It continued even though twenty-five of the world's top AI researchers asked it to stop, citing its much higher error rates when identifying dark-skinned women than light-skinned men.[150]

And it persisted even after Buolamwini testified before the US Congress House Oversight and Reform Facial Recognition hearing in 2019, where she laid out the problems in response to New York Congresswoman Alexandria Ocasio-Cortez's questions.[151]

> **Ocasio-Cortez:** Ms. Buolamwini, I heard your opening statement, and we saw that these algorithms are effective to different degrees. So are they most effective on women?
> **Buolamwini:** No.
> **Ocasio-Cortez:** Are they most effective on people of color?
> **Buolamwini:** Absolutely not.
> **Ocasio-Cortez:** Are they most effective on people of different gender expressions?
> **Buolamwini:** No, in fact, they exclude them.
> **Ocasio-Cortez:** So what demographic is it mostly effective on?
> **Buolamwini:** White men.
> **Ocasio-Cortez:** And who are the primary engineers and designers of these algorithms?
> **Buolamwini:** Definitely white men.

Only after George Floyd, a black man, was killed in May 2020 by Derek Chauvin, a white police officer, sparking protests across the United States as well as in other countries, did Amazon, IBM and Microsoft announce that they would stop selling their facial recognition technology to US law enforcement.[152] (Google had already said in January 2020 that it would not sell facial recognition to law enforcement agencies.) So here we have a rare and public example of where and when to draw the line: outside a convenience store in South Minneapolis, Minnesota on 25 May 2020.

When some of the most powerful technology companies in the world are voluntarily restricting the use of their products and asking the government for regulation, it is time to pay attention.[153] At first, Amazon said it would restrain itself for only one year – the time that Jeff Bezos, then its CEO, thought it would take for the US Congress to regulate the technology. Was he being naive or optimistic?

Since Floyd's killing, Congress has introduced two bills – the Facial Recognition and Biometric Technology Moratorium Act and the Fourth Amendment Is Not For Sale Act – but the US legislative process is slow by design.[154] In the best-case scenario it will be a while before we see any movement at the federal level, and in the worst-case scenario, nothing will happen at all. In May 2021 Amazon drew the line again, announcing that it was extending its moratorium on the police use of facial recognition *indefinitely*.[155]

By contrast, action is happening quickly at the city and state level. More than a dozen cities and states have banned police use of facial recognition technology or passed moratoriums. A few have also restricted private companies' use of it.

All this legislating shows that we are moving on from the earlier argument that we can 'solve' the problem of facial recognition technology by making it more accurate. To do that, we need more data about skin tones, genders, sexualities and ages, of course. The more and diverse the data, the better!

That argument misunderstands the problem. Even if we solve the *technical* problem of facial recognition technology's inaccuracy, the *political* problem persists. Facial recognition technology is about more than data and algorithms: it is about privacy, civil liberties and the balance of power between consumers and companies and between citizens and the state.

For example, in 2021 Clare Garvie, a senior associate at the Center for Privacy and Technology at Georgetown Law, explained some of the other reasons for concern regarding US police use of facial recognition technology to the US current affairs show *60 Minutes*:[156]

> **Anderson Cooper:** Is facial recognition being misused by police?
>
> **Clare Garvie:** In the absence of rules, what – I mean, what's misuse? There are very few rules in most places around how the technology's used.
>
> **Cooper [voiceover]:** It turns out there aren't well-established national guidelines and it's up to states, cities and local law enforcement agencies to decide how they use the technology: who can run facial recognition searches, what kind of formal training, if any, is needed, and what kind of images can be used in a search. In a 2019 report, Garvie found surprising practices at some police departments, including the editing of suspects' facial features before running the photos in a search.
>
> **Garvie:** Most photos that police are – are dealing with understandably are partially obscured. *Police departments say, 'No worries. Just cut and paste someone else's features, someone else's chin and mouth into that photo before submitting it.'*
>
> **Cooper:** *But, that's, like, half of somebody's face.*
>
> **Garvie:** *I agree. If we think about this, face recognition is considered a biometric, like fingerprinting. It's unique to the individual. So how can you swap out people's eyes and expect to still get a good result?* [Emphasis added]

Regulating police use of facial recognition technology will not, on its own, prevent such abuses: we must also regulate the private sector. If we do not, we will leave a loophole for law enforcement to exploit – just as it is already is by making enthusiastic use of Amazon's Ring doorbell video surveillance system.[157]

To investigate the use of facial recognition technology in the United States is like lifting a rock and finding all sorts of creepy crawlies underneath:

- many US training datasets are a problem;
- many US algorithms are a problem;
- the role of the US government, and especially that of the US military, in developing and expanding this technology is a problem for Americans and non-Americans; and
- the patchwork of regulation is a problem, as this makes US citizens' experience of privacy and civil liberties, which are protected by the Bill of Rights, conditional upon where they live.

This matters for anyone who lives in or visits the United States. It also makes the UK position on facial recognition technology very, very awkward.

United Kingdom
While writing this book I have had the opportunity to discuss facial recognition technology on BBC national radio with three police officers – two from the London Metropolitan police and one from South Wales Police. All of them insisted that their facial recognition technology works perfectly well and does not have the same problems with inaccuracy that researchers in the United States have found repeatedly.[158]

I was intrigued. Had the United Kingdom procured superior facial recognition technology and somehow concealed it from its American friends? If so, surely the Americans would have noticed by now and switched suppliers?

Alas, no.

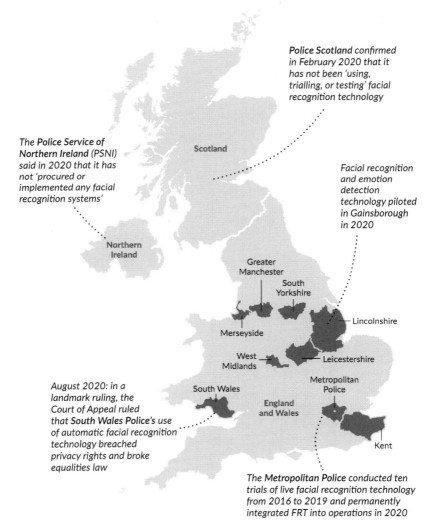

● UK police forces known to have used facial recognition technology

Police Scotland confirmed in February 2020 that it has not been 'using, trialling, or testing' facial recognition technology

*The **Police Service of Northern Ireland** (PSNI) said in 2020 that it has not 'procured or implemented any facial recognition systems'*

Facial recognition and emotion detection technology piloted in Gainsborough in 2020

Scotland

Northern Ireland

Greater Manchester

South Yorkshire

Lincolnshire

Merseyside

West Midlands

Leicestershire

Metropolitan Police

*August 2020: in a landmark ruling, the Court of Appeal ruled that **South Wales Police's** use of automatic facial recognition technology breached privacy rights and broke equalities law*

South Wales

England and Wales

Kent

*The **Metropolitan Police** conducted ten trials of live facial recognition technology from 2016 to 2019 and permanently integrated FRT into operations in 2020*

Figure 13. UK police forces known to have used facial recognition technology.

In May 2018 *The Independent* newspaper published the results of its freedom of information request into facial recognition trials being conducted by the Met and the South Wales Police: of the 104 alerts that were triggered in the former's trials, only two were accurate – a false positive rate of 98% – while the latter's results showed accuracy in fewer than 10% of cases.[159] That same year, Big Brother Watch published an investigation based on fifty freedom of information requests that found that, on average, 95% of UK police 'matches' wrongly identified innocent people.[160] It also named the Met as the worst offender for incorrect matches.

In 2019 Professor Peter Fussey and Dr Daragh Murray published their 'Independent report of the London Metropolitan Police Service's trial of live facial recognition technology'. They found that across six deployments, only eight of the forty-two matches were correct (19.05%).[161] Their report was commissioned *by the Met*, and yet when I discussed it on BBC Radio 5 Live in January 2020 with presenter Stephen Nolan and Metropolitan Police Federation Chairman Ken Marsh, the following exchange took place:[162]

> **Stephen Nolan:** Well, hold on a minute, in terms of that piloting, you're talking about the trials which have taken place in 10 occasions and locations such as Stratford's Westfield Shopping Centre and the West End of London.
> **Ken Marsh:** Yes.
> **Nolan:** The Met said, when it tested the system using police officers, the results suggested that 70% of so-called wanted suspects would be identified while only one in 1,000 people generated a false alert. That's what the Met's saying. But there was then an independent review – not the Met's trials, but an independent review of six of these deployments – that used different tech methodology, and that found that only eight of the forty-two matches were correct!
> **Marsh:** Independent by who?

Nolan: So in other words, that thirty-four of them were incorrect!

Marsh: Independent by who?

Nolan: Well the point is, it wasn't done by the Met –

Marsh: Independent by who?

Stephanie Hare: Well, I mean, I'm holding the 'Independent report of the London Metropolitan Police Service's trial of live facial recognition technology' in my hands right now and that is done by the University of Essex. Its authors are Professor Pete Fussey and Dr Daragh Murray. It was done in July 2019. And this report says that we need to have a legal framework, we need to have regulation, and we need to test this technology far more, and that it doesn't work! It's not reliable and accurate enough.[163]

It was all very odd.

We simply cannot ignore or gloss over facial recognition technology's inaccuracy problems. Being misidentified by the police can have grave consequences, even when facial recognition technology is not involved. Most of us will remember what happened in 2005 when Met police officers stormed a train in London and shot dead Jean Charles de Menezes, a twenty-seven-year-old Brazilian electrician who they had incorrectly identified, by human error, as a suspected terrorist.[164] De Menezes lost his life; his family and friends lost a loved one; and the Met spent years fighting legal action before eventually paying out hundreds of thousands of pounds in compensation and fines.

That is an extreme and rare example of what can happen from misidentification. It is harder to judge the degree of harm from the following example, which I reported on in 2018 for the BBC World Service and which involved facial recognition technology as well as human error.

Silkie Carlo, director of Big Brother Watch: Recently, the [London] Metropolitan Police used facial recognition outside a shopping mall. They had what they thought was an identification come up. It was a young black male. He was apparently, according to observers, stopped, searched, asked to show his ID, emptying his pockets in front of members of the public, and he was misidentified. Big Brother Watch's investigation found that 98% of the Metropolitan Police's so-called matches were actually misidentifications, and yet these people are then ending up on police databases.

Stephanie Hare: Detective Superintendent (DS) Bernie Galopin of the Metropolitan Police says there are safeguards in place as to how the technology is being tested.

DS Bernie Galopin: So we had a deployment recently during the summer. And there was a false positive. It resulted in a male being stopped. A false positive is an incorrect identification. On this particular day we had a very small rate of false positives. But in relation to this individual, I saw the image of the individual and I also saw the image of the person on the watchlist. In my opinion, there were real similarities. But this is where we have the safeguard in place, where we had an officer, who stopped the individual, who spoke to them about the fact that a positive alert had been generated, and after conducting some checks, realized that this was not the person who was identified through the technology. I myself spoke to the individual afterwards. He was very positive about the experience. For me, that provides the reassurance that we are looking at the ethical considerations. It's technology that's assisting the police here, not police assisting in the technology. *About 99% of the people who walk through were not falsely identified. We've not had a single complaint from a member of the public.* [Emphasis added]

Should we find it reassuring that no one complained? Not necessarily. Not everyone feels comfortable complaining about the police stopping them and demanding that they show their ID and empty their pockets due to a misidentification.

We do not all have the same experience when we have to deal with the police or the Home Office.[165] For instance, a report from Her Majesty's Inspectorate of Constabulary and Fire and Rescue Services (HMICFRS) in February 2021 showed that black people are far more likely than white people to be stopped by the police in England and Wales, and twice as likely in London. Black people are also more likely to have force used on them, to be handcuffed, to have a spit and bite guard used on them, and to have their mugshot published. 'Over 35 years on from the introduction of stop and search legislation, no [police] force fully understands the impact of the use of these powers,' the HMICFRS report said. 'Disproportionality persists and no [police] force can satisfactorily explain why.'

Meanwhile, there's little point in complaining about the police using facial recognition technology if doing so can lead to a fine, as it did for one man who was walking down the street in East London in 2019.[166] BBC television reporters were there to film the Met's trial of facial recognition technology and captured what happened:[167]

Silkie Carlo, director of Big Brother Watch, who was there to witness the trial: What's your suspicion?
Plainclothes officer: The fact that he's walked past clearly masking his face from [facial] recognition.
Silkie Carlo: I would do the same.
Plainclothes officer: He's covered his face.
Silkie Carlo: I would do the same.
Plainclothes officer: It gives us grounds to stop him –.
Silkie Carlo: No it doesn't!
Man who was fined: The chap told me down the road, he said, 'They've got facial recognition.' So I walked past like that [demonstrates pulling his top up to cover his face] – it's a cold day as well. As soon as I've done that, the police officer's asked me to come to him. So I've got me back up. I

said to him, '**** off', basically. I said, 'I don't want me face showing on anything.' If I want to cover me face, I'll cover me face! It's not for them to tell me not to cover me face. I've got a £90 fine now. Here you go, look at that [holds the fine up to the camera]. Thanks, lads. £90. Well done.[168]

The BBC Click report went viral on social media, showing that misidentification is not the only danger of facial recognition technology: it also changes the balance of power between citizens and the state, weakening our privacy and civil liberties and exposing us to the risk of police overreach. It also helped to raise public awareness of the fact that 'no legislation exists that explicitly authorises police use of live facial recognition technology', as Peter Fussey and Daragh Murray wrote in their 2020 analysis of the Met's trials of the tool.[169]

Furthermore, our national data regulators 'do not have explicit authorisation to limit live facial recognition technology'.[170] Until Parliament passes a proper legislative framework, the police and the private sector can use facial recognition technology on us without our knowledge or consent and there's little anyone can do to stop it short of taking legal action.[171]

This is not ideal. As Professor Paul Wiles, then the United Kingdom's Biometrics Commissioner, told the BBC World Service in 2018:

We have now got a whole new generation of biometrics that are being experimented with or deployed by the police but they're also being used by the commercial sector. That really needs a legislative framework. The government's biometric strategy does not propose to do that.[172]

And here is Tony Porter, then the Surveillance Camera Commissioner, speaking to *The Times* in 2019:

Ministers must act over the explosion of spy technology...
[It is] unacceptable that no law has been introduced to control how intrusive technologies such as facial recognition were used.[173]

Steve Wood, Deputy Information Commissioner, was also worried, as he told the House of Commons:

There are potentially thousands of custody images being held with no clear basis in law or justification for the ongoing retention.[174]

Wood was referring to the fact that police databases in the United Kingdom contain the facial data of innocent people alongside mugshots of suspected and convicted criminals. A High Court ruled in 2012 that they must not do this, but the Home Office, which controls the police, ignored this until 2017, when it devised a workaround: if we think our face is wrongfully in a police database, we can request to have it deleted.[175]

The problem is that most of us do not know we have this right, there is little evidence that citizens are taking advantage of it, and the police can refuse our request anyway.[176] This means that when the police scan our faces, they may find us in their database *even if we have done nothing wrong*. This requires us to prove our innocence, which violates one of the core tenets of a liberal democracy: we are presumed innocent unless proven guilty in a court of law.

Darren Jones, Member of Parliament for Bristol North West, brought the regulators' concerns to the attention of the House of Commons in 2019:

I do not want to keep saying that everybody agrees with me, because that would be a little uncomfortable, but there is no denying that the Biometrics Commissioner, the Surveillance

Camera Commissioner and the Information Commissioner's Office have all said exactly the same thing – [the UK] biometrics strategy is not fit for purpose and needs to be done again.[177]

In response, the Commons Select Committee recommended a moratorium on the use of facial recognition technology until a new legislative framework could be established:

This Report highlights growing evidence from respected, independent bodies that the lack of legislation surrounding the use of automatic facial recognition has called the legal basis of the trials [of automated facial recognition technology] into question.

The Committee calls on the Government to issue a moratorium on the current use of facial recognition technology and no further trials should take place until a legislative framework has been introduced and guidance on trial protocols, and an oversight and evaluation system, has been established.[178]

And how did the governments of Theresa May and Boris Johnson respond to this recommendation? They ignored it.

Finally, in August 2021, the Home Office published a proposed code of practice governing police use of live facial recognition in England and Wales. However, Tony Porter, the former Surveillance Camera Commissioner, told the BBC this was 'bare bones':

I don't think it provides much guidance to law enforcement, I don't really [think] it provides a great deal of guidance to the public as to how the technology will be deployed.[179]

This lack of guidance prevents the public from knowing just how much our civil liberties now vary across the United Kingdom.

HOW CIVIL LIBERTIES VARY ACROSS THE UNITED KINGDOM
Today the United Kingdom is in a similar position to the United States on a constitutional matter: our experience of policing, privacy and civil liberties now depends on our postcode.

England
The Met is not the only police force in England to use facial recognition technology.[180] Gordon's Wine Bar in London, which created a facial recognition technology company called Facewatch, told The *Financial Times* that by 2016 it was working with more than twenty police forces.[181] However, some police forces – such as West Midlands Police and Kent Police – have refused pressure from the Home Office to use the systems.[182]

Wales
South Wales Police received a £2.6 million grant from the government in 2017 to test live facial recognition technology, mainly at large events such as sports matches and concerts but also at protest marches and in shopping centres.[183]

The force used the technology in 2017 at a peaceful protest in Cardiff city centre, which prompted Ed Bridges, whose facial image was captured, to file a lawsuit. In August 2020 the High Court ruled that this was 'unlawful' and that there was 'no clear guidance on where [the technology] could be used and who could be put on a watchlist, a data protection impact assessment was deficient and the force did not take reasonable steps to find out if the software had a racial or gender bias'.[184]

Bridges told the BBC that he wants the UK government to act so that 'discriminatory technology like this is banned for good'. He added: 'We have policing by consent in this country. The police need to have the support of the public in what they do and my concern is that by using a technology that is discriminatory and not being used in accordance with the law, that actually the police then lose the support of the public. And that's not in anyone's interest.'

In December 2021 police in South Wales and Gwent announced that they will become 'the first in the United Kingdom to identify wanted individuals in real time through a new facial recognition app on their mobile phones. During the testing phase, which will last for 3 months, the app – known as Operator Initiated Facial Recognition – will be used initially by 70 officers from South Wales Police and Gwent Police.'[185]

Scotland

In February 2020 Police Scotland confirmed that it had not been 'using, trialling, or testing' facial recognition technology. It also postponed plans to use it until at least 2026 after the justice sub-committee of the Scottish Parliament warned that there was no justification for using it, that to do so would be a 'radical departure' from Britain's 'policing by consent' model, that the technology is 'known to discriminate against females and those from black, Asian and ethnic minority communities', and that any police use of this technology would need to be *provided for in legislation* and meet human rights and data protection requirements' [emphasis added].[186]

Northern Ireland

The Police Service of Northern Ireland has confirmed that as of 2018 it had not 'trialled, procured or implemented any facial recognition systems' but that it 'intends to include the use of facial recognition technology within its Biometrics programme. There is no current project or date set yet for initiation of a project.'[187] In its responses to a freedom of information request in 2020, it said that it has not 'procured or implemented any facial recognition systems'.[188]

Our analysis of how facial recognition technology relates to power in the United Kingdom offers the following list of takeaways.

- Our regulators are either unwilling or unable to stop the proliferation of facial recognition technology.[189] This is only likely to worsen now that the roles of Surveillance Camera Commissioner and Biometrics Commissioner have been merged into one and the new holder, Fraser Sampson, who is a former officer and solicitor for the police, has said that facial recognition technology should not be banned and its use should be a matter of police discretion.[190]
- The High Court ruling against South Wales Police was enough to stop that police force from using the technology, but it did not stop the Met, which in 2020 announced that it was integrating the technology into its operations permanently.
- No action on facial recognition technology will be coming from Westminster so long as Boris Johnson's government retains its eighty-seat majority in the House of Commons. The next election is due to be held on 2 May 2024.
- As of this writing, a private member's bill submitted by Liberal Democrat peer Timothy Clement-Jones in 2019 calling for a moratorium on and review of facial recognition technology still hasn't been heard.[191]
- Scotland shows that it is possible to stop the police use of facial recognition technology at the level of devolved government. Even so, the devolved governments of Wales and Northern Ireland are unlikely to follow suit.

Our experience of policing, privacy and civil liberties differs depending on where we live, but there is more to our experience of facial recognition technology than political philosophy. To understand how, we must view it through the lens of aesthetics.

AESTHETICS: WHAT IS OUR EXPERIENCE OF FACIAL RECOGNITION TECHNOLOGY?

Nobody knows how many surveillance cameras there are operating in the United Kingdom today. In 2018 the official estimate was 6 million cameras (for a population of 66 million people, or one camera for every eleven people), but Tony Porter, who served as the Surveillance Camera Commissioner until March 2021, believes the number is higher.[192] London was the second most monitored city in the world in 2019, after Beijing, with an estimated 420,000 CCTV cameras operating in and around the city, and we must remember that surveillance cameras go well beyond CCTV: they include 'body worn video, automatic number plate recognition, vehicle borne cameras and unmanned aerial vehicles (drones)'.[193]

Pauline Norstrom, former chairman of the British Security Industry Association, a trade body representing the United Kingdom's private security industry, also thinks the number is higher. 'The number of video-surveillance cameras has been growing in the UK at an annual rate of 10% a year, to around 10 million cameras in the country,' she said in 2020.[194]

It would take technical effort and financial investment to move from CCTV to facial recognition technology, but it is possible. There are no laws to prevent it – indeed, it is already happening.

For instance, in 2019 the *Financial Times* reported that convenience stores and grocery stores such as Budgens, Tesco, Sainsbury's and Marks & Spencer have cameras that were either 'already, or soon to be, capable of facial recognition'. It noted that millions of patients in NHS hospitals across the country are exposed to surveillance from drones, body worn video and automated facial recognition. It reported that Facewatch, the facial recognition technology created by the owner of Gordon's Wine Bar in London, was being tested by 'a major UK supermarket chain, several major events venues and a prison', and had been launched with fifty retailers in the area around London's Victoria railway station, including Co-op Food.[195]

In 2019 Big Brother Watch found that facial recognition technology was being used by 'major property developers, shopping centres, museums, conference centres and casinos'.[196] That same year the *Evening Standard* reported that the City of London Corporation had approved plans for some of London's top cultural and retail destinations – including the Barbican arts complex, Hay's Galleria and Liberty department store – to use facial recognition technology.[197]

To live or travel in the United Kingdom is to experience constant surveillance. Increasingly, we are not passive in this: many of us are spying on each other, which creates a different kind of experience. Sales of Amazon Ring doorbells – data from which can be shared with the police unless we actively opt out of that happening – are rising. A *Sunday Times* investigation from 2019 showed that several police forces have formed partnerships with Amazon to create a public–private surveillance network.[198]

Ring doorbells offer video surveillance only, not facial recognition technology, but this could change. Many other companies already offer such a product, with Nest, which is owned by Google, being one example.[199] Amazon is thinking about it. In 2018 it filed two patents for facial recognition technology. The following year it told US Senator Ed Markey that facial recognition technology was a 'contemplated, but unreleased, feature' of its home security cameras but that 'there are no plans to coordinate that feature with its law enforcement partnerships'.[200]

This is just some of the surveillance that we know about. Occasionally we discover surveillance we did not know about. This alters our sense of reality: what kind of a world are we really living in?

For example, in 2019 the *Financial Times* reported that a private landlord had been using facial recognition technology *in secret* for twenty-two months in London's King's Cross, working with the Met police and the British Transport Police.[201] This area is frequented by thousands of people every day, and the landlord's cameras would have been able to analyse the faces of people travelling via the Eurostar terminal, those frequenting the offices of *The Guardian*, Google and DeepMind, as well as anyone visiting the area's restaurants, bars and clubs. They would have captured me and everyone who attended my public lecture in 2019 at the CogX festival of AI in King's Cross, where I warned of the dangers of facial recognition technology and called on Parliament to consider a moratorium on its use.[202]

As of this writing, this whole episode remains a mystery. At first the police denied involvement, only to do a U-turn a few days later and admit that, yes, they were involved.[203] The Mayor of London and members of the London Assembly claimed that they knew nothing about it. The Information Commissioner's Office opened an investigation. More than two years later, it has still not decided whether this surveillance was legal.[204] It has not even said why the investigation is taking so long, what it needs to make a decision, or when we might expect one. Such a lack of accountability affects our lived experience: it suggests that our system of governance is broken and that we are powerless to do anything about it.

Calls to rethink our use of facial recognition technology have come from an unexpected stakeholder: our intelligence agencies. In 2021 these warned that local authorities' use of 'smart city technology', which includes CCTV and facial recognition, could be used by Beijing for espionage, surveillance or collection of sensitive data.[205] They want limits or even bans on certain vendors, many of which have already been designated as national security threats by the US government.[206]

This, too, may affect our lived experience. It is not just our own authorities and private companies that we need to worry about spying on us: it is foreign powers, too. In 2019 *The Intercept* reported that cameras made by Hikvision, which is 42% owned by the Chinese government, is on the US export controls list

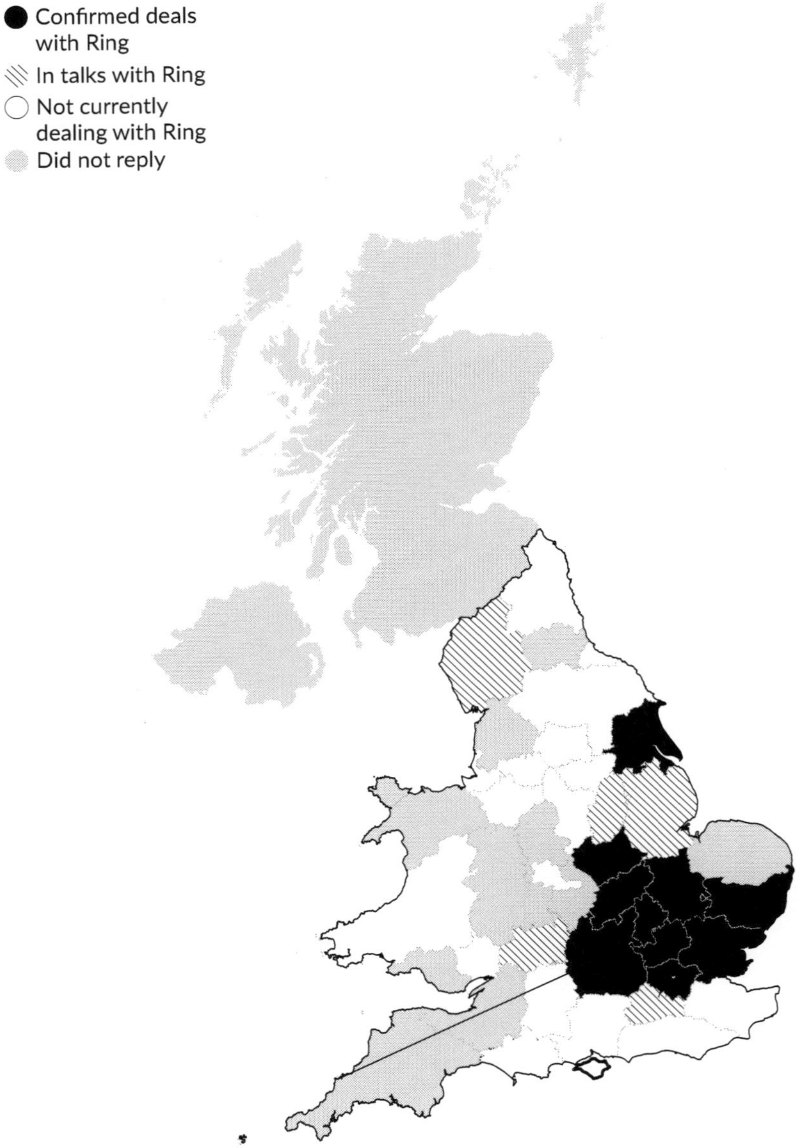

Figure 14. Partnerships between Amazon Ring and police forces in England and Wales as of 2019, adapted with permission from the *Sunday Times*.

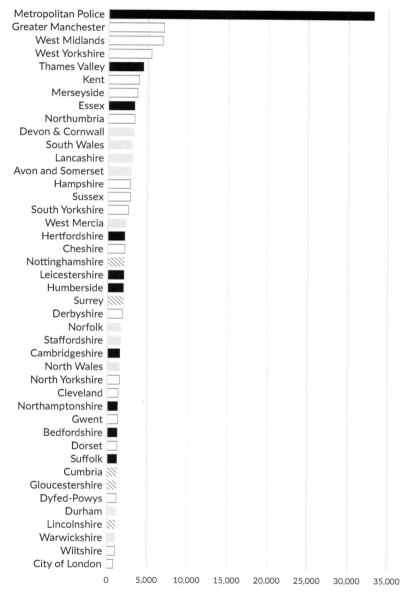

Figure 15. The number of police officers in each police force (full-time equivalent, 2020), adapted with permission from the *Sunday Times*.

because it poses a 'significant' risk to US national security.[207] It also supplies the surveillance cameras used in the mass internment centres in Xinjiang province in Western China, and its systems are installed in airports and on 'the British parliamentary estate' as well as being supplied 'to police, hospitals, schools, and universities'.[208] Admiral Lord Alan West, a member of the House of Lords, even alleged that new Hikvision cameras had been installed in London's Portcullis House, where 'more than 200 members of parliament and 400 of their staff work'.[209]

More than 1.2 million surveillance systems used throughout the United Kingdom are already provided by some of those Chinese vendors.[210] According to Thomson Reuters:

> Freedom of information requests filed in late 2020 by digital rights researcher Samuel Woodhams with all 32 London councils and the next 20 largest UK city councils found about two-thirds owned technology made by two Chinese companies [Hikvision and Dahua, both of which are prohibited in the United States on national security grounds] accused of links to the repression of Uyghurs in China.[211]

Still, there is some good news. According to Thomson Reuters, all of the UK councils that bought Chinese technology have said that they are not currently deploying facial recognition systems.[212] However, nothing would prevent them from doing so in the future. Nor are they obligated to inform the public proactively if they do so – we will be dependent on researchers submitting freedom of information requests. How does this affect our experience of living in the United Kingdom, knowing that our local authorities could choose to use surveillance technology on us without even informing us, much less obtaining our consent?

Meanwhile we are left with a dilemma. We have already paid for a lot of surveillance technology from China. Ripping it out will cost us more money, and so will replacing it. Yet we must weigh those costs against the cost of not ripping it out and replacing it. This is all money that could be spent on other things that improve our experience of life in our cities. Political choices do not only have political consequences. They also shape our life experience (aesthetics) and our sense of reality (metaphysics). Whether we think this is good or bad is a question of ethics.

ETHICS: IS FACIAL RECOGNITION TECHNOLOGY A GOOD THING OR A BAD THING?

In June 2021 fifty global asset management firms overseeing more than $4.5 trillion between them said that they would ask the companies they invest in to ensure that facial recognition technology is developed 'in an ethical way, with the right regulation and oversight'.[213] But not everyone agrees that it is possible to develop facial recognition technology in an ethical way. 'Often the problem is that the topic itself is unethical,' the independent technology ethicist Gemma Galdón-Clavell told *The Guardian* in 2020. She went on:

> Some topics encourage partners to develop biometric tech that can work from afar, and so consent is not possible – this is what concerns me. *If we're talking about developing technology that people don't know is being used, how can you make that ethical?*[214] [Emphasis added]

Ethics has already permeated our analysis of facial recognition technology in this chapter. We learned that it is not just a biometrics technology – it is also a technology for surveillance and control. We traced its origins as a tool of policing, racism, colonial administration and border control, and we looked at its connections with the debunked pseudosciences of physiognomy and phrenology. We mapped its most common uses and found that almost all of them pose a high risk of harm. We explored how facial recognition technology is a tool of power, and one that is more likely to misidentify people with darker skin, particularly women.

We have learned that in the United States it has led to wrongful arrests; that police have sometimes tampered with facial images; that people may be denied their right to a fair trial by not being informed that they were identified using facial recognition technology; that some of the world's most powerful technology companies are restraining themselves from selling facial recognition technology to the police; and that a number of cities and states have stopped waiting for the federal government to regulate or ban police use of this tool and have simply done so themselves.

We learned that in the United Kingdom researchers have also found that facial recognition technology misidentifies people, and that we need more research into bias and discrimination in policing in general, and into predictive policing algorithms in particular. We have learned how the High Court issued a landmark ruling that the police use of facial recognition technology was 'unlawful' but that this has not deterred the Met from permanently integrating it into its operations. We have seen that the three main data regulators* think that our biometrics strategy is not fit for purpose and needs to be done again; that the Science and Technology Committee in the House of Commons recommends a moratorium on the use of facial recognition technology until a proper legislative framework is put in place; that a bill was put forth by a Liberal Democrat peer for a moratorium and a review nearly two years ago but it has not yet been heard; that the likelihood that the police will use facial recognition technology on us depends on where we live; and that the private sector's use of the technology is a free-for-all.

---

* Except for Professor Fraser Sampson, who used to be an officer and a solicitor for the police, who thinks that facial recognition should not be banned and should be left to police discretion.

We learned that China is using facial recognition technology to surveil and control its population and to commit what the UK government and some of its closest allies have called 'genocide'.

We explored how living in the United Kingdom is to live in a goldfish bowl, constantly watched by CCTV cameras installed by the police, by private companies and increasingly by each other, thanks to the growing number of us who are purchasing Amazon Ring doorbells and other private cameras. We learned that CCTV and, increasingly, facial recognition technology are being used on us while we shop, while we visit museums, while we visit a bar or simply when we visit certain London neighbourhoods. We learned that our regulators either cannot or will not do anything to stop this, and that our intelligence agencies are worried that this data – our face, our identity – could be taken by a foreign power using the surveillance systems that our government procured (paid for by our taxes) to watch us.

As we ask whether facial recognition technology is good or bad, we might refine our question further.

- For whom is it good or bad?
- Under what conditions is it good or bad?
- What unintended consequences might there be to using it?

Facial recognition technology is not a subject we can shrug off, saying 'to each their own', because it affects all of us – and some of us more than others.

We need a lens that allows us to assess the impact of this technology on our society. Utilitarianism, which asks how we can maximize benefits and minimize harm, is a concept pioneered by the British philosopher Jeremy Bentham. He also gave the world the idea of the panopticon: a prison with a watchtower in the centre from which the warden can see any inmate at any time. By contrast, the inmates can never be certain whether they are being watched – and so logically they should assume that they are being watched at all times (see Figure 16).

Figure 16. Exterior and interior of the Presidio
Modelo in Cuba, built as a panopticon.

Bentham got the idea for the panopticon from his brother,
Samuel, who devised it while working in Russia, where he
needed a way to supervise his workforce.[215] He grasped that the
panopticon could be used for all sorts of purposes, not just prisons and labour management. It could be 'a new mode of *obtaining power of mind over mind*' [emphasis added].[216]

The French philosopher Michel Foucault agreed, exploring
Bentham's panopticon both as a physical system and as a met-
aphor for how surveillance and social control exist throughout
society. 'Is it surprising,' he asked in *Discipline and Punish*, 'that
prisons resemble factories, schools, barracks, hospitals, which
all resemble prisons?'[217]

Both Foucault and Bentham would probably be stunned by how we have realized the concept of the panopticon – and extended it. Why build a physical panopticon to watch us when it is so much easier to create a digital one that can identify and track us without us even knowing?

CONCLUSION

We have not completed our panopticon yet. There is still time for us to change course, but only if we decide *now* what our ethics are in the face of biometrics and surveillance technologies, so that we can craft a proper legislative framework that is fit for purpose and honours our centuries-long tradition of privacy and civil liberties. As we do this, we can learn from other parts of the world.

First, we will need to watch developments in the European Union, which has already demonstrated its power in setting global data standards and which proposed much stricter regulation of all 'remote biometric identification systems' in 2021.[218]

The European Commission has published a proposal – the AI Act – to regulate AI, in which the police use of live facial recognition technology in public places could be 'prohibited in principle', although exceptional uses could be allowed such as 'finding abducted children, stopping imminent terrorist threats and locating suspects of certain crimes, ranging from fraud to murder'. This could only be done with prior approval from a judge or a national authority.[219]

As things stand, this text is inadequate, because it fails to regulate private use of facial recognition technology and leaves loopholes for both governments and companies to exploit.[220] There is a risk that judges or national authorities will simply rubber stamp any requests from the police or security services, as occurred in the United States after the attacks of 11 September 2001, when secret courts set up under the Foreign Intelligence Surveillance Act rubber-stamped law enforcement applications to tap tech companies' data.[221] We need a proper legislative framework, not judges and national authorities applying a very broad interpretation of existing anti-terrorism laws.[222]

There are signs that some European elected representatives want to go further. In October 2021 the European Parliament adopted a non-binding resolution calling for a ban on police use of facial recognition technology in public places unless it is to fight 'serious' crime, such as kidnappings and terrorism, *and* a ban on private facial recognition databases. This followed the United Nations Human Rights chief Michele Bachelet's call in September 2021 for a moratorium on the sale of and use of artificial intelligence technology – including facial recognition – that poses human rights risks.[223]

Second, we cannot ignore what is happening in the United States, where a growing number of cities, counties and states are either regulating facial recognition technology or banning it altogether.

Some of the most powerful technology companies in the world are based in the United States and they are refraining from selling their systems to law enforcement for fear of civil liberties and human rights abuses. In June 2021 senators reintroduced legislation to ban government use of biometric technology, including facial recognition.[224] Again, by focusing solely on regulating government use of facial recognition technology, lawmakers create a loophole for law enforcement and the security services to exploit: they can simply access private use of facial recognition technology.

Third, we must not turn away from the role of facial recognition technology in China as part of a system of social control, the crushing of political expression in Hong Kong, and the repression of the Uyghurs on ethnic and religious grounds.

The UK government has already taken considerable steps to condemn these activities. Now it must confront the risks posed by our vast and growing surveillance network here at home, and not just in terms of our suppliers. We must face our failure to protect privacy and civil liberties *equally* throughout the country so that we are a *United* Kingdom, not one in which our freedoms depend on where we happen to live.

Facial recognition technology has made it pointless to protest against a national ID card by threatening to eat it, masticate it or cut it up with a penknife, as Prime Minister Boris Johnson once did. We already have an ID card that works nationally and internationally: our face. What we do not have is a proper legislative framework to protect that 'ID card' from being used and abused by the police, companies, foreign powers and each other. Until we do, we need an immediate moratorium on medium- and high-risk uses of facial recognition technology so that we can decide how to regulate it – or even to ban it.[225]

Johnson, with his strong and longstanding opposition to the introduction of a national identity card in the United Kingdom, would seem a natural fit for bringing forward such legislation to Parliament, thereby protecting all who live here from this surveillance technology. Yet in 2021 he did a U-turn on this question. First his government unveiled plans to require photo ID at elections throughout the entire United Kingdom, not just in Northern Ireland, even though we have almost no voter fraud.[226] Then, in July 2021, he warned that he had plans to require a 'vaccine passport' for domestic use in England during the coronavirus pandemic. This would amount to a national ID card – one that contained not only our personal information but our Covid certification status, accessed by using facial verification technology on our smartphone.[227] By the end of the year, all four nations of the United Kingdom had adopted this as a legal requirement in some settings (the details of which will be discussed in the next chapter).

That's another thing about asking where we draw the line: we may not have to draw it only once. We may have to draw it over and over again, as values, facts and situations evolve. As we do so, we create new maps of power, reflecting new realities of where the police, the private sector and individuals can use our data and where they cannot, transforming the relationships between citizens and the state, between consumers and companies, and between humans and technology.

Table 2. Attention, lawmakers: facial recognition technology examples ranked by risk of harm and need for regulation.

Example of how facial recognition technology is used	Risk of harm and thus need to regulate
To unlock our smartphone	Low
To control borders and immigration	Low
To analyse our physical health	*Medium*
To pay with our face	*Medium* (**high** for children)
To tag/be tagged on social media	*Medium*
To monitor workers (e.g. Uber and Amazon drivers)	**High**
To enter a building	**High**
To access humanitarian assistance	**High**
To identify a person in a crowd	**High**
As a tool of foreign policy	**High**
As a tool with which to wage war	**High**
To access government services (e.g. Aadhaar)	**High**
To apply online for a passport	**High**
To classify faces by ethnicity, race, sexual orientation, political affiliation or emotional state	**High**

Chapter 4

Pandemic? There's an app for that

Boris Johnson ✓
@BorisJohnson
⚑ United Kingdom government official

This is going to be a fantastic year for Britain.

10:18 AM · Jan 2, 2020 · TweetDeck

Figure 17. New Year's greetings from UK
Prime Minister Boris Johnson, 2020.

As I write, a novel coronavirus is flying through the air, trying to kill us.*

---

* I'm paying homage to George Orwell here, who in 1940 conveyed the terror and tension of the Blitz in the first line of his essay 'The lion and the unicorn: socialism and the English genius': 'As I write, highly civilized human beings are flying overhead, trying to kill me.'

Strictly speaking, SARS-CoV-2 is not *trying* to kill us. It is *trying* to find a host whose cells it can use as raw material to recreate itself – and in the process of doing that, it can kill us. Like every other virus, it must strike a balance: if it kills too many hosts too quickly, it will not be able to keep spreading and will burn out. The most successful viruses are those that can spread the furthest and last the longest, so sometimes they mutate into more contagious variants, as this one has done several times. This requires a more varied transmission strategy: one that allows enough hosts to live long enough to spread the virus.

SARS-CoV-2 is crafty. Some of us catch it and never know because we experience no symptoms (a status called asymptomatic). We feel fine. We continue going about our daily lives, all while spreading the virus. Some of us fall ill and experience symptoms ranging from those of a cold to a bad case of flu, while others are left with 'long Covid' – damage lasting weeks, months or even a year or more.

And some of us die a terrible death. While this group consists mainly of the elderly and those with pre-existing health conditions, it also includes some who are young, healthy and have no underlying health conditions. As of 15 December 2021, the World Health Organization had recorded 5,318,216 deaths from Covid-19 worldwide – of which 146,627 were in the United Kingdom.[1]

In addition to the traditional disease-control measures we have deployed to stop the virus from spreading – such as handwashing, the wearing of face masks, social distancing, working from home, lockdowns, contact tracing, quarantine and vaccines – we have also explored several new digital health tools, such as immunity passports, exposure notification apps, Quick Response (QR) code check-in programmes and vaccine passports. This chapter will assess the new digital health tools that the United Kingdom[*] considered, created and/or rejected

---

[*] Since health is a matter for devolved government in the United Kingdom, this chapter will focus mainly on the experience in England, where the majority of the UK population lives (including the author).

between March 2020, when we went into our first lockdown, and December 2021.[2] It will consider if any of these tools were effective and whether any of them have earned a place in our toolbox for future pandemic preparedness.

IMMUNITY PASSPORTS

When England went into its first lockdown, from 23 March 2020 to 4 July 2020, many governments considered an immunity passport as a way of reopening their countries.[3] This was to be based on tests showing that either we currently had the virus or that we had already had it and recovered. Some countries, such as Chile, announced plans to create this kind of scheme in April 2020, only to halt them a month later when it became clear that there was still so much we did not know about the virus. At that time we did not know whether having been infected with SARS-CoV-2 conferred immunity, and it was also unclear whether any such immunity was temporary or permanent.

To further complicate matters, it quickly became apparent that immunity passports risked creating perverse incentives. For instance, people might try to become infected with the virus in the hope that they would survive it, obtain an immunity passport and regain their freedom. Meanwhile, those who 'failed' to get infected with the virus – perhaps because they were obeying the government's orders to 'stay home, save lives, protect the NHS' – would remain confined to their homes.

This created the risk of a two-tier society as well as raising questions about privacy and civil liberties with respect to workers' rights, employers' responsibility for the health and safety of their workers, and policing. For example, employers could make immunity passports a condition of returning to work after lockdowns – again, this would have incentivized people to get infected and take their chances. Also, police might have demanded to see a person's immunity passport when they were out in public. After all, from October 2020 they were empowered to access contact tracing data from human contact tracers in order to learn who has been told to self-isolate, and to then fine anyone who was failing to do so £1,000 – and up to £10,000 for repeat offenders.[4]

As a prospect, immunity passports were considered and dismissed fairly quickly. However, they represented an important dress rehearsal for the debates that followed about some of the other experimental technologies we deployed, such as exposure notification apps, QR code scanning, and vaccine passports for domestic use.

EXPOSURE NOTIFICATION APPS

'It turns out that it is very, very challenging to get people to use a Covid app,' Professor James Larus told the *New York Times* in December 2020. 'We went into it thinking that of course people would want to use this, and we have been very surprised.'[5]

Larus was one of several computer scientists who worked with Apple, Google and public health officials in the spring of 2020 to create a new and experimental technology called an Exposure Notification Application Programming Interface (API). This uses Bluetooth technology to record the digital IDs of nearby phones.[6] While these digital IDs are called 'contacts', they are not contacts: they are indicators of proximity between two smartphone users. They are also randomly generated and frequently changed, so that user privacy is preserved – and traditional contact tracing is not possible.[7] As a result, national public health agencies could use the API to create their own exposure notification apps to warn smartphone users that they had been exposed to someone who had tested positive for Covid-19, and the API would protect users' privacy by keeping their data on their smartphone.

Many countries adopted the Google Apple Exposure Notification model from the outset, but there were three exceptions:

Germany, France and the United Kingdom. These three countries wanted to build an app based on a centralized model, as this would give their public health authorities far more data that could be used to understand the spread of the virus and it would support their human contact tracing teams. By contrast, a decentralized model offers more privacy but that also means less data, limiting the ability of public health authorities to understand the spread of the virus. Moreover, French authorities objected to the two Silicon Valley giants dictating the terms of a pandemic public health response and therefore decided to build their own national app.[8]

In April 2020 Germany did a U-turn following the publication of an open letter from nearly 300 academics who argued in favour of the Apple– Google API and warned that exposure notification apps built on a centralized model risked 'a form of government or private sector surveillance that would catastrophically hamper trust in and acceptance of such an application by society-at-large'.[9] This left only France and the United Kingdom to continue building their tools based on a centralized model through the spring of 2020.

The debate about which data model to use was important and took up considerable bandwidth, but it also skipped over a more fundamental question: could an exposure notification app help to break the virus's chain of transmission? If so, would it deliver a better return on investment than other public health mitigations that might be more effective, more equitable or more affordable?

To answer such questions, we needed to know what exposure notification apps could and could not do. Throughout the pandemic the media has often referred to exposure notification apps as 'contact tracing apps'. This is incorrect and misleading, so let us start by getting clear on terminology.

Contact tracing is done by contact tracers, who are usually public health officials but can also be volunteers or people who are hired and trained. Contact tracers help us to map out everyone we have come into contact with since becoming infected. Then they contact those people and map out *their* contacts, and so on. The aim of contact tracing is to map out the chain of transmission – and break it.

Contact tracing is privacy invasive, and this can cause problems: not everyone wants to share who they have been in contact with and under what circumstances. Even when people are happy to disclose their contacts, not everyone can remember everyone they came into contact with or knows their names and details. Contact tracing is a race against time, and since the novel coronavirus and its variants transmit quickly, some virologists and public health officials thought it was worth trying to develop a digital health tool that could alert people to an exposure to infection more quickly.[10]

Exposure notification apps are a new and experimental digital health tool that many countries around the world developed during the pandemic. They have a specific and limited function: to alert anyone who has downloaded the app if they have been exposed to someone who has tested positive for the virus. The app may also advise you to self-isolate for a specified period of time and invite you to request a test. By design, the person receiving the alert will never know who exposed them or when the exposure occurred, and neither will anyone else. Unless they alert the public health authorities, no one will know of their potential exposure. Unless they take a test and – if it comes back positive – enter the result into the app, none of the people they potentially exposed, and who also have the app, will receive an exposure notification alert.

Exposure notification apps are an example of technology ethics in action since their creation and use requires us to make decisions about values: privacy and patient confidentiality versus public health surveillance. As Susan Landau, professor of cybersecurity at the Fletcher School, observes, 'under this model privacy and confidentiality win but public health authorities such as contact tracers and epidemiologists lose out', since exposure notification apps do not provide information to track the spread of the virus.[11]

Google, Apple and the various technologists who worked on exposure notification apps were mindful of how data gathered by apps can be used and abused. Since use of the apps was voluntary, they reasoned that preserving privacy and confidentiality might reassure people and encourage uptake.

Such concerns were especially relevant in the United Kingdom for a number of reasons.

- First, the country has experienced many alarming data breaches, which have alerted the public to the risks of data collection.
- Second, Matt Hancock, who served as Health Secretary until June 2021, was forced to overhaul his 'Matt Hancock MP' app in 2018 after some of his constituents complained that it was ignoring their privacy settings and harvesting their personal information, including photos, friend details, check-ins and contact information.[12]
- Third, in 2019 researchers discovered that a smartphone app created by the Home Office for European citizens to apply to live and work in the United Kingdom after Brexit had vulnerabilities that could allow access to phone numbers, addresses and passport details, including facial scans.[13]
- Finally, Hancock appointed Dido Harding to lead NHS Test and Trace despite the fact that when she was CEO of TalkTalk in 2015, the company failed to implement even the most basic cybersecurity, allowing the theft of the data of 157,000 customers, much of it unencrypted.[14]

Despite this poor track record, or perhaps because of it, the United Kingdom created a number of privacy-preserving exposure notification apps that some, though by no means all, smartphone users could use if they so wished. Here are some of the lessons we learned.

Lesson 1. Exposure notification apps will not work without accurate and fast test turnarounds

To determine whether an exposure notification app could help break the chain of transmission, we needed to know more about the virus and the holy trinity of 'test–trace–isolate', which public health authorities worldwide maintained was needed to suppress the virus. (In the United Kingdom, this holy trinity is an unholy pentagram of 'find–test–trace–isolate–support'.)

The Johns Hopkins Bloomberg School of Public Health offered anyone in the world a free, six-hour online course for contact tracing. The course showed how contact tracing works on a *human* level – something that it is essential to understand before considering how it might work with an app.[15] For example:

- the incubation period of the infection is 2–14 days, but mostly within 3–5 days;
- people are most infectious with this virus on the first day that they show symptoms;
- although no one will show symptoms sooner than two days after being infected, they are still infectious for forty-eight hours before they first show symptoms – a status known as pre-symptomatic;
- the 'window of opportunity' to break the chain of transmission is therefore in the first seventy-two hours of infection.

Since an infected person will not show symptoms for at least forty-eight hours, they need to be able to arrange a test and get the results back within twenty-four hours of first starting to display symptoms. Ideally, they would be tested before they even start showing symptoms, which could occur if they were alerted within the first forty-eight hours of an exposure.

Twenty-four hours from test request to result is a tight turnaround. It requires a country to have a well-functioning testing capability. The United Kingdom did not have such a capability when it first started designing its exposure notification app, and as of June 2021 – a month before England opened up and eased most pandemic restrictions – it still did not.[16]

On the contrary, the country's testing capacity has been beset with problems (although it has improved during the writing of this book). For example, it was endangered by supply chain problems, which meant we risked running out of the ingredients and equipment needed to perform the tests. Then the government had to investigate why the entire testing operation for Public Health England was being run using Microsoft Excel, the technical limitations of which meant that 15,841 positive coronavirus test cases were not passed on to human contact tracers – putting an estimated 50,000 contacts at risk of exposure to the virus.[17] Economists Thiemo Fetzer and Thomas Graeber estimate that this led to 125,000 further cases and 1,500 deaths.[18]

A twenty-four-hour test turnaround would have helped to make the United Kingdom's exposure notification apps more effective, but only if at least three other conditions were met.

- First, we needed to have enough test kits available so that anyone in need could request it, in contrast to our earlier policy of rationing tests due to shortages.
- Second, we needed to test asymptomatic people, not just people with symptoms, as was the policy for several months, again due to test shortages.[19]
- Third, we needed to ensure that the technology linking test results and national contact tracing capabilities worked to avoid any further repeats of the unfortunate Excel episode.

Devi Sridhar, professor of global health at Edinburgh University, summarized things well when she said that we need 'a robust system for testing, tracing and isolating, where test results are returned within 24 hours, at least 80% of people's contacts are reached, and there is a high adherence to a rule of 14 days' isolation for those exposed to the virus'.[20]

Lesson 2. Exposure notification apps are of little use if people cannot afford to self-isolate

Just as an exposure notification app will fail without proper testing, it will also fail if people cannot or will not self-isolate when instructed to do so. It is insufficient to simply notify people of an exposure: the whole test–trace–isolate system must work.

This brings us to the question of compliance. In September 2020 a study claimed that exposure notification apps can make a difference even with adoption rates as low as 15%. However, this assumed that there was a 90% likelihood that people who were notified would begin a fourteen-day quarantine, when in reality the opposite is true: most people in England (82%) do not fully self-isolate even when they have tested positive for coronavirus, according to the government's Scientific Advisory Group for Emergencies (SAGE).[21]

Unless compliance increased, SAGE argued, mass testing and contact tracing and exposure notification apps would be of little use to break the chain of transmission.[22] To improve compliance, the government applied a carrot-and-stick approach in England.

The stick was fines ranging from £1,000 for first-time offenders to £10,000 for repeat offenders (this is for England and Wales; Scotland and Northern Ireland have their own fines).[23] Police officers can check that we are complying based on 'local intelligence'.[24] This measure was not without controversy: Professor Lucy Yardley, a member of SAGE, warned that fines may make people afraid to report symptoms and seek a test, because this could lead to their family members and work colleagues also being asked to self-isolate when they cannot afford to do so.[25]

The fines only applied if we were told to self-isolate by a *human* contact tracer.[26] A senior source at the Department of Health also confirmed that the exposure notification app's 'ping' was 'advisory', because the authorities could not legally enforce something that cannot be proved (such as a potential exposure to the virus).[27]

This was just as well, as the exposure notification app was beset by teething problems – all too common after an initial launch – such as an inability to upload positive test results, disappearing alerts (which cause anxiety among some users) and 'ghost' alerts that confused some users about whether they needed to self-isolate or not.[28]

The carrot was financial support. At first, this amounted to only £182 (or £13 a day) for the fourteen-day isolation period, and it was only trialled in North West England.[29] This was far from adequate, so it was hardly surprising that many people ignored instructions to self-isolate. Indeed, as SAGE's scientists warned the government:

> The poorest people are the least likely to self-isolate because they cannot afford to stay home. Self-reported ability to self-isolate or quarantine is three times lower in those with incomes less than £20,000 or savings of less than £100.

In September 2020 the government increased the support package for self-isolating to £500 (or £35.71 per day) for the fourteen-day period.[30] But again this was far from sufficient, and by January 2021 compliance was still failing. According to the *Financial Times*, a government survey showed that only 17% of people with symptoms were coming forward to get a test because they feared that a positive result would stop them from working (and thus earning), while the applications of three-quarters of the people who had applied for the £500 lump sum payment to support self-isolation were rejected.[31]

Bear in mind that until December 2020, the possibility of financial support was only an option for people who had been instructed to self-isolate by *human contact tracers*. This meant that anyone who been using *the app* since it was launched in September and was advised to self-isolate was not entitled to apply for support.[32] And even after the government expanded support to low-income workers pinged by the app, most who applied were turned down.[33]

In February 2021 *The Guardian* published a leaked report from the Joint Biosecurity Centre that concluded that 'unmet financial needs' meant that people in poorer areas were less able to self-isolate, and that this, along with low wages, cramped housing and the failures of the NHS Test and Trace scheme, had led to the 'stubbornly high' coronavirus rates in England's most deprived communities.[34]

On 1 April 2021 the *British Medical Journal* reported that just '18% of those with coronavirus symptoms said they had requested a test, while 43% with symptoms in the previous seven days adhered to full self-isolation'.[35] People less likely to self-isolate included men, younger people, people with young children, people from working-class backgrounds and key workers.[36] Their reasons for failing to self-isolate included needing to go to the shops or work, needing medical care (for something other than Covid-19), having caring responsibilities, to take exercise, to meet others, and having mild symptoms.[37]

By June 2021 *The Guardian* reported that councils in England were refusing more than six out of every ten requests for the £500 self-isolation support, with the proportion rising to more than 90% in some parts of the country.[38] That month, Health Secretary Matt Hancock explained that this was because the government worried that people might try to 'game' the system:

The challenge that we had with that proposal is the extent to which it might be gamed. Because after all, a contact gives test and trace their contacts. You wouldn't want a situation where if you tested positive you could then enlist your entire friendship network to get a £500 payment.[39]

Gaming the system is a legitimate concern, which is why it is so striking that Hancock and his colleagues in government were so relaxed about giving taxpayer money to *their* friends during the pandemic. We are not just talking about the odd £500, either: according to Transparency International UK, '27 PPE or testing contracts worth *£2.1 billion* were awarded to firms with connections to the Conservative Party' [emphasis added] – many of them without tender.[40]

An exposure notification app is of limited use against such politics. The reality of the pandemic is this: many people needed financial support to self-isolate; without it, they had to keep working to live, which meant they risked spreading the virus.[41]

Lesson 3. Design for interoperability as early as possible

The United Kingdom built five contact tracing apps that did not share the same name and were not interoperable until November 2020 – nine months after work first began on creating them. The apps in question were 'Beat COVID Gibraltar', 'Jersey COVID Alert', 'Protect Scotland', Northern Ireland's 'StopCOVID NI' and England and Wales's 'NHS Covid-19' (not to be confused with the NHS app, which before the pandemic was for booking appointments and since 2021 has included our Covid vaccine passport). This reflects the political system in the United Kingdom, where health is a matter for devolved governments.

Although the England and Wales app covered 90% of the UK population, the different names of the apps initially created confusion. Furthermore, the lack of interoperability created real problems for people who live and work on the borders (e.g. between Scotland and England) or for those who travel between the different nations because these users had to remember to download different apps and to switch between them. This introduced friction into the user experience and made it more likely that some people would fail to use the apps correctly.

To coordinate better, the devolved nations of the United Kingdom had to do further work, costing more money, resources and time. In October 2020 the National Cyber Security Centre announced that it was beginning work to achieve interoperability across the Common Travel Area (which comprises the United Kingdom plus Jersey, Guernsey, the Isle of Man and the Republic of Ireland), and afterwards would expand into Europe and then further afield.[42] The following month it announced that all five UK apps were interoperable.[43]

On the one hand, this situation is unsurprising: 'agile' software delivery focuses on getting the basics out of the gate and then building iteratively on top of those basics. On the other hand, countries that only needed to build one app for their entire population may have been at an advantage, as they avoided this extra work, saving them time and resources.

Until well into 2021 none of the UK apps were interoperable with the apps of any other countries. This is not surprising: none of the European countries designed their apps to work together from the outset – a point of reflection for the future. On 29 September 2020 Ireland, Germany and Italy announced that they would link their apps to a common system to facilitate safer travel, with Austria, the Czech Republic, Denmark, Estonia, Spain, Latvia and the Netherlands to be linked as well by the end of October.[44] Users of those apps would not have to download anything else; rather, they would be able to use their own country's app as they travelled.[45] This interoperability evolved further from November 2020 when it was announced that several vaccines had been developed, so that by the summer season of 2021 various EU countries recognized each other's vaccine certification.

Lesson 4. Seek feedback early and often, including from 'citizen science' and 'critical friends'

Millions of people in the United Kingdom began tracking their Covid-19 symptoms from April 2020 using the ZOE COVID app, which was created and launched in five days by doctors and scientists from King's College London working with the health science company ZOE. This app identified some Covid symptoms before the public health authorities did, such as loss of smell and loss of taste, as well as new symptoms in fully vaccinated people who still became infected, such as sneezing.[46] In a similar tale of community sourcing of information, members of the gay community in the United States discovered in July 2021 that it was possible for fully vaccinated people to still be infected by the Delta variant and alerted the Centers for Disease Control and Prevention.[47] Both of these examples of 'citizen science' made important contributions to our understanding of the virus and led to updated public health guidance.

As with citizen science, the role of 'critical friend' existed long before the pandemic, as it is good practice for anyone building new tools and technologies to invite constructive criticism throughout the design and launch process. In the United Kingdom a number of critical friends studied the various exposure notification apps and offered ideas on how to improve them – some privately, some publicly, some by invitation and some on their own initiative.[48] These critiques probably contributed to the government's June 2020 announcement that it would follow Germany and do a U-turn, abandoning the centralized model and going instead for the decentralized (privacy-preserving) model built on the Apple–Google API.[49]

It takes courage to abandon an idea on which we have spent time and resources and which politicians and the media have hyped up to the public. The sooner we can decide that something is not working, the sooner we can redeploy our efforts to something that stands a greater chance of success. By comparison, it took the French government another four months to admit that it too would have to redesign its app – a waste of precious time and resources.[50]

EXPECTATION MANAGEMENT

The government has been under pressure from the media and the public to improve its pandemic response, and every failure and setback is seized upon and held up for a thrashing, no doubt partly because our prime minister insists on describing our every effort as 'world-beating'. This sets technology up to fail. Far better to remind people both that exposure notification apps are an experimental technology and that while they might help break the chain of transmission, they also might not.

Had politicians acknowledged that fact from the outset – telling the public that we were going to try something new and inviting everyone to give feedback in the hope that together we could create something that helped to stop the spread of the virus – we might have had a more constructive experience. Moreover, even if the app is deemed to have delivered a poor return on investment or to have 'failed', the media and politicians should not frame this as a wasted effort: we will have gained important data and experience as to why it did not work, which could be useful not only during this pandemic but for our public health preparedness in the future.

In September 2020 the government launched version 2.0 of its app for England and Wales. While privacy and confidentiality were now preserved by design, another problem remained: digital exclusion. Many smartphone users could not download the app because their phone was too old and could not support the operating software. However, this problem was not limited to the new England and Wales app. Rather, it applies to every app that runs on the Apple–Google API: an estimated 2 billion phones worldwide cannot access it.[51]

This put a new technological spin on an old ethical dilemma.

- Is it ever acceptable to exclude people from healthcare?
- If so, who is it acceptable to exclude and on what grounds can they be excluded?
- Who decides who can be excluded?
- Do excluded persons have any recourse or must they accept their fate?

The virus does not care what kind of phone we have, and we should not have to pay hundreds of pounds to upgrade from our otherwise perfectly functional smartphones in order to download an app that all taxpayers helped to fund. We are all at risk here, so why are only some of us being offered this extra protection? NHS stands for *National* Health Service, not 'health service for people who already own a particular model of smartphone or can afford to upgrade'.

This is a tricky problem to solve. On the one hand, the limited accessibility of the NHS Covid-19 apps is down to the technology, not the government: the Apple–Google API on which the UK apps are built requires a minimum threshold of technology on the phone in order to operate.[52] But on the other hand, politics and ethics did still play a part: an update was announced in December 2020 that would, in theory, have allowed people with older iPhones to get the exposure notification app[53] (Android phone owners were still out of luck) but it was then announced in February 2021 that the plan to extend the app to older iPhone models had been 'put on the back burner in order to prioritise other features'.[54]

Our understanding of how the virus spreads improved over time, which presented another design challenge for exposure notification apps: some infected people appear to transmit the virus more than others. Germany's top virologist, Christian Dros-ten, who advised Chancellor Angela Merkel, explained the dif-ference in approach to *Bloomberg Businessweek*, who reported that:

> His thoughts on testing have evolved. If health officials get overwhelmed in the coming months, he argues, they should stop trying to capture every case of Covid. Most people only pass it on to one other person anyway, posing little systemic risk. But occasionally, someone spreads it more widely, cre-ating a cluster of infections that really drives the pandemic. To avoid another lockdown, he says, Germany must focus on identifying and preventing those clusters.[55]

Zeynep Tufekci, a professor of sociology who writes about technology and the pandemic, agreed, writing in *The Atlantic* that:

> Multiple studies from the beginning have suggested that as few as 10 to 20 percent of infected people may be responsi-ble for as much as 80 to 90 percent of transmission, and that many people barely transmit it...
>
> To fight a super-spreading disease effectively, policy mak-ers need to figure out why super-spreading happens, and they need to understand how it affects everything, including our contact-tracing methods and our testing regimes.[56]

Critical friends also highlighted the need to consider the downsides of accessibility, such as mission creep, which occurs when something created for one purpose is used for another, often with little consideration for whether this is a good idea.

For example, wearable devices issued by the government to every single person would improve accessibility. This was one reason why Singapore moved to take this step in July 2020.[57] However, Singapore's wearable is less privacy-preserving than the Apple– Google API used in the country's smartphone exposure notification app, as it requires users to give their national ID numbers and phone numbers. Moreover, such a wearable device could easily shift from a mission of alerting people to possible virus exposure to become a tool for law enforcement to arrest anyone who violates a self-isolation order – or to track them for other purposes. As the University College London lecturer Michael Veale explained to the BBC, all a government would have to do is fit Bluetooth sensors to public spaces in order to be able to identify anyone with a wearable device.[58] 'All you have to do is install physical infrastructure in the world,' he said, 'and the data that is collecting can be mapped back to Singapore ID numbers.'[59]

Such warnings proved prophetic. In January 2021 Singapore's government confirmed that the police would be able to use data gathered from its exposure notification app *and* its wearable notification device in criminal investigations, even though it had previously explicitly ruled this out and said that the data would be used only for virus tracking.[60] By that stage, nearly 80% of Singapore's residents were already signed up to the programme.

The Singapore experience is a powerful argument for *privacy by design*: if you want to be certain that the police, the government or anyone else cannot access data from a system you are creating, make that core to your design from the outset. Do not assume you can fix this problem afterwards – it will almost certainly be too late.

Lesson 5. The metrics that matter are deaths, hospital admissions and vaccine uptake – not the exposure app's adoption rate

In April 2020 researchers at the University of Oxford published a study suggesting that the *adoption rate* required for exposure notification tracing apps to work would be 80% of smartphone users.[61] But even with a lower adoption rate, the app may still help to break the chain of transmission.

In May 2021 a peer-reviewed article was published in *Nature* that examined the app during the period 24 September–20 December 2020, when the app was downloaded onto 21 million unique devices (62%) but only actively used by 16.5 million people.[62] Using a modelling approach, the researchers estimated that the number of cases averted was 284,000 while the estimated number of deaths avoided was 4,200; using a statistical approach, the respective numbers were 594,000 and 8,700.[63]

That sounds significant, but as BBC technology correspondent Rory Cellan-Jones noted, the researchers could only provide estimates because 'the private nature of the app means there are some key questions that can't be answered'. Questions such as the following:

- 'How many people obeyed the ping on their phone telling them to self-isolate?'
- 'How many of them had been separately contacted by the manual track and trace operation anyway?'
- 'How many of those alerts were false positives or negatives, meaning people who were not at risk were told to self-isolate while the reverse was true of others?'
- 'How many people grew bored with the app and switched off Bluetooth or uninstalled it?'[64]

In the *Nature* article the researchers were cautious. 'Theoretical evidence has supported this new public health intervention but its epidemiological impact has remained uncertain,' they wrote in their abstract, repeating in the article that 'digital tracing is a novel public health measure, with unknown epidemiological impact'.[65] Moreover, after sharing their encouraging estimates, they emphasized that the app is just one part of a system of non-pharmaceutical interventions: it is not a substitute for social distancing, the wearing of face masks or human contact tracing.[66]

On 12 May 2021 Christophe Fraser, one of the *Nature* article's authors and a professor at the University of Oxford, explained on Twitter that:

> The app has similar accuracy to manual recall of contacts, but is reaching more contacts per person. Of course, manual tracing has broader reach: both approaches have pros and cons, you need both; but it's useful to see them side by side.

And he was more bullish by July, telling *The Times*:

> We have shown through analysis that high usage of the app does make a difference to a community.
>
> If you compare two local authorities side by side that are at a very similar stage of the epidemic and have a similar socio-economic make up and so on – having more app users is one of the factors that results in lower infection rates in those communities, so it does make a difference.[67]

By this point in the pandemic, though, questions about the exposure app in England and Wales had evolved from how well it might work to whether it might work a bit *too* well.

After missing its first target date for lifting most restrictions on June 21, the government did so (in England) on 19 July – 'Freedom Day', as it was dubbed by some of the media. This was a milestone in more ways than one. England's vaccination rate had reduced the number of deaths and hospitalizations to a level that was politically acceptable and operationally manageable for the public health authorities. Breaking the chain of transmission and keeping cases down was no longer the priority.

In July and August the number of cases soared, and unsurprisingly, so did the number of smartphone users pinged by the app and advised to self-isolate. This 'pingdemic' occurred nowhere else in the United Kingdom or in other countries that used exposure notification apps.[68] In response to complaints of worker shortages and fears that people would ignore the alerts or delete the app, the public health authorities adjusted the app on 2 August so that fewer people would be pinged.[69] Again unsurprisingly, the pingdemic eased.

All this drama became much less relevant from 16 August, when the government changed its policy so that fully vaccinated people were no longer advised to self-isolate if pinged by the app. They were instead asked to take a PCR test, as it is still possible for fully vaccinated people to get infected and to infect others.[70]

While the exposure notification apps were still up and running at the end of September, the government's focus was not so much on the number of cases as it was on the number of hospitalizations. It judged these to be at a manageable level. However, it reserved the right to reinstate restrictions, including lockdowns, if hospitalizations ever become too high and the NHS comes under intolerable strain.[71] When this became the case in December 2021, the exposure notification apps were not mentioned as part of the government's 'Plan B' measures.

QUICK RESPONSE (QR) CODE CHECK-IN

Another digital tool that the government created in 2020 was NHS QR code check-in, which aimed to alert people of exposure to the virus linked to a particular location.

QR code technology was created in 1994 by Masahiro Hara, chief engineer at the Japanese automotive components company Denso Wave, who got the idea while playing Go, a board game that originated in China in the second millennium BCE. 'I used to play Go on my lunch break,' Hara explained in 2020. 'One day, while arranging the black and white pieces on the grid, it hit me that it represented a straightforward way of conveying information. It was a eureka moment.'[72]

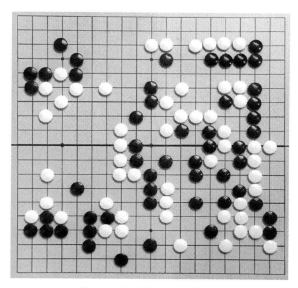

Figure 18. A game of Go.

A QR code can embed 200 times more information than a standard barcode, and it allows a scanner to find and interpret the code's information twenty times faster than previous matrix codes.[73] Before the pandemic it was used for digital boarding passes, mobile payments and to convey information at tourist spots, in shops and on gravestones. During the pandemic it has been used for digital menus in restaurants and digital check-ins at venues to facilitate Covid-19 exposure notification alerts.

In the United Kingdom, owners of any venue could go to a government website and create a QR code for them to display. Visitors then scanned the code upon their arrival at the venue using their exposure notification app.[74] If someone who had been at that location later tested positive for the virus, anyone who had scanned the code into their phone within the window of infection would receive an exposure notification and an instruction to isolate for fourteen days and request a test. An example QR code can be seen in Figure 19.

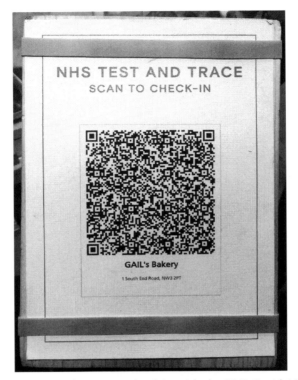

Figure 19. QR code scan to check in with the NHS Covid-19 App.

Businesses were required to display a QR code. However, customers were not required to scan those codes, since not all customers have smartphones or could download the app. Scanning, therefore, was done on a voluntary basis. If customers did not scan the code, they were supposed to record their name and contact details, usually on a piece of paper or in a notebook. This was neither private nor secure: anyone could look at the paper/ notebook and take the information. Some women reported being contacted by men for romantic/sexual purposes using the contact information that the women had given to venues for NHS Test and Trace.[75] Some people (not just women) protected themselves from such data insecurity by giving fake information, rendering the system less effective in the process.

Despite these problems, from October 2020 onwards many people participated in QR code scanning. Then, in March 2021 Sky News got hold of a leaked confidential report showing that data from hundreds of millions of check-ins had barely been used by NHS Test and Trace, 'potentially leading to the spread of the virus'.[76] Only 284 alerts were sent, for 276 venues, despite more than 100 million venue check-ins in total using the app. Furthermore, when the data was used, public health officials instructed businesses to contact the exposed individuals directly rather than going through NHS human contact tracers, thereby breaching data protection law and exposing businesses to lawsuits.[77]

This was not ideal.

Things improved from April 2021, when more alerts started being sent, reaching a peak of 1,992 alerts from venues in England and Wales in the week ending 21 July 2021.[78] However, the numbers began to decline after that, because from 19 July the government no longer required businesses to ask customers and visitors to check in.

VACCINE PASSPORTS FOR DOMESTIC USE

In November 2020, after the announcement that several vaccines had been developed and would be rolled out with immediate effect, something curious happened in the United Kingdom: Prime Minister Boris Johnson and several members of his cabinet began briefing the media that they had no plans to introduce vaccine passports for domestic use (also known as Covid passes or Covid certification).[79] This was a sure sign that introducing such a measure was on the cards, as were the ministers' denials, which were worthy of the best performances in London's West End.

- Cabinet Minister Michael Gove told BBC Breakfast: 'Let's not get ahead of ourselves, that's not the plan. What we want to do is to make sure that we can get vaccines effectively rolled out.'[80]

- Health Secretary Matt Hancock said: 'I'm not attracted to the idea of vaccine passports here. We are not a papers-carrying country.'[81]
- Vaccine Minister Nadhim Zahawi ruled it out, arguing that they could be 'discriminatory' since it is not compulsory for people to get the vaccine.[82]

Meanwhile the government was funding eight pilot schemes to build vaccine passports, including one that uses facial recognition technology and another that uses fingerprint vein technology.[83] By January 2021 it was trialling vaccine passports for domestic use on a smartphone app in two local authorities.[84]

The public U-turn occurred the following month, when *The Times* reported that a vaccine passport could be added to the NHS app – not the exposure notification app that we discussed earlier in this chapter but a different app that allows users to book appointments and see their medical records.[85] Now the government planned to expand the NHS app's existing functionality to show if users had received a vaccine, recently tested negative for the virus, or had 'natural immunity', which was established by having tested positive in the previous six months.

At no point did the government give an explanation for its U-turn. The people of the United Kingdom glossed over this to focus on an apparently more important matter: would we need a Covid-19 vaccine passport to go to the pub?[86]

Prime Minister Boris Johnson kept changing his answers to questions about vaccine passports. On 15 February he said: 'What I don't think we will have in this country is, as it were, vaccination passports to allow you to go to, say, the pub or something like that.'[87] Less than six weeks later, though, on 24 March, he said that it might be up to pub landlords (not Parliament!) to decide if vaccine certification would be needed to gain entrance to their pubs.[88] The following day he suggested that it might only be possible to implement a vaccine passports scheme once all adults had been offered a jab, which would not be until some time in September.[89] In the end, he instructed Michael Gove to lead a review and report back by 21 June – the date on which the government planned to fully reopen the country.

A national debate ensued.

Most of the public, the media and Parliament did not object to the idea of requiring a vaccine passport for international travel, as there are already precedents for this (e.g. a requirement to show proof of vaccination against yellow fever).[90] By contrast, requiring a vaccine passport to go to the pub, a restaurant, the cinema, a concert, the theatre, a festival, a sporting match or the workplace would be an altogether different matter. As discussed in the previous chapter, the United Kingdom does not require anyone living here to carry ID. Vaccine passports for domestic use would amount to a national ID card by stealth – and therefore to a shift in the balance of power between the state and citizens, employers and workers – as well as being an untested public health tool. As such, it would likely require primary legislation.[91]

Writing in *The Telegraph* in April 2021, Gove said it was time to explore vaccine passports for domestic use – something he called 'Covid certification'.[92] He did not seem particularly persuaded by the scheme, explaining that he believed there were some places where we should never have to demonstrate our Covid status, such as the supermarket, the pharmacist and a GP's surgery. The virus, of course, does not care about locations, so it seemed from the outset that requiring vaccine passports for domestic use was *not* about breaking the chain of transmission. So what *was* it about?

'We need to ask where else it would be wrong to require certification,' Gove wrote. He went on:

Where should the lines be drawn to help protect freedoms, respect privacy, promote equality and get us back to normality? And how can we ensure our approach is proportionate and time-limited? Those are the questions we need to ask in the days ahead.[93] [Emphasis added]

He concluded by asking for the public to share their views on where the line should be drawn – and in return he got an earful.

Some people opposed vaccine passports for domestic use on civil liberties grounds.[94] A group of eight psychologists and behavioural scientists, most of whom were on the government's Scientific Pandemic Insights Group on Behaviours (SPI-B), warned that vaccine passports could encourage people to deliberately get infected in order to be allowed to re-enter society and also abandon wearing masks, social distancing and other mitigations – an echo of the debate we had had about immunity passports in early 2020.[95] Peter Yapp, a former deputy director of GCHQ's National Cyber Security Centre, warned that the app could become a honeypot for hackers.[96] The Royal College of General Practitioners warned that it could be discriminatory against those who did not have smartphones or were less tech-savvy.[97] The Equality and Human Rights Commission said that it risked creating a 'two-tier society' between those who had had their two jabs and those who had not, such as the young (who had not even been offered a jab at that point and who would not be fully vaccinated until September), as well as groups among whom vaccine take-up was lower, such as migrants, those from minority ethnic backgrounds, and poorer socioeconomic groups.[98] The hospitality industry largely opposed the proposed scheme because it did not want to have to police customers and turn them away.[99]

Sir Jonathan Montgomery – who chaired the ethics advisory board for NHSX on its contact tracing app – then gave a masterclass of logical reasoning when he explained his concerns on BBC Radio 4's *Today* programme:

The first is the scientific one – *does it work?* – and that all depends on this information about risk of transmission.

The second is a timing issue. We need to reopen the economy as quickly as it's safe to do so, and *vaccine passports are not going to be useful until people have had their second vaccine. So it's not something that's going to solve the problem for summer 2021, because* even on the fantastic achievements that we've had, *the population that is going to use nightclubs is not going to have had its two vaccinations until at least the autumn, and we need everything open before then if we can.*

And then the third question is *who gets excluded by this.* So if you haven't been able to get the vaccine, you get excluded. If you for whatever reason are not appropriate to have the vaccine, or you have objections to using the vaccine, you get excluded. And *those things are likely not to be evenly distributed across society.*

Now if this was the only way of getting the clubs open, then we might trade off the intrusions into privacy, but *if there are other ways of doing it, then we probably wouldn't want to have private information shared unnecessarily.*[100] [Emphasis added]

Several Conservative MPs had already expressed reservations about the idea, but when they learned that the digital agency behind the development of the government's exposure notification app had started work on vaccine passports, 'lots of people on the [briefing] call said, "Oh no!",' according to one MP.[101] Forty-one of them ended up publicly signalling their opposition.

These Conservatives were joined by Sir Keir Starmer, the Labour opposition leader, who described the idea as 'un-British'. Again, this was a purely political argument that took no account of the scientific question of whether they actually work. Eventually he did find his way there:

My instinct is that ... [if] we get the virus properly under control, the death rates are near zero, hospital admissions very, very low, that the British instinct in those circumstances will be against vaccine passports.[102]

Sir Ed Davey, the Liberal Democrat leader, also appealed to politics as well as ethics rather than asking if vaccine passports would help prevent death and reduce hospitalizations or drive down transmission: 'As well as burdening struggling pubs with extra cost,' he said, 'The idea that businesses can voluntarily bar certain customers, who may not even have been offered a vaccine, is deeply illiberal.'

The cross-party opposition may have been enough to reject any proposal for vaccine passports for domestic use had it ever come to a vote.[103] The government therefore introduced them without one.

By the end of May it had scrapped its plans to require vaccine passports for domestic use. In June it maintained a discreet silence when the Public Administration and Constitutional Affairs Committee called on the government to abandon its 'unjustified' plans, arguing that it had not made the scientific case.[104] In July it published its review into vaccine passports for domestic use and again confirmed that it would not require them:

> The review has concluded that the Government will not mandate the use of COVID-status certification as a condition of entry for visitors to any setting at the present time. While the review concluded that there would be a public health benefit, the impacts are judged to be disproportionate to the public health benefit at this stage of the pandemic.[105]

Meanwhile, it deployed the vaccine passports for domestic use that it had been building all along, making them available to venues to use by choice rather than requiring them by law.[106] The Premier League began using them on a voluntary basis, as did some nightclubs.[107]

The government did this even though some of the companies it had paid to build the vaccine passports were on record saying that governments could 'redeploy this effort into a national citizen ID programme' (Entrust), that the government should include facial verification checks on the door at venues (iProov), and that digital IDs for Covid should be 'recognisable to law enforcement and other agencies to prove a person is immune' (Onifido).[108]

On 22 June 2021, people in England began reporting on social media that a new feature – Event Trial – had suddenly appeared on the NHS app, and it allowed users to show their vaccine status.[109] The app also now said that the Covid vaccine passport could be used for 'any domestic arrangements going forward'.

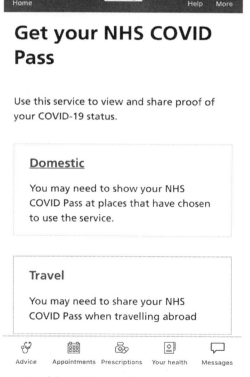

Figure 20. 'The NHS Covid Pass will be dependent on ... domestic arrangements going forward.'

In July Prime Minister Johnson announced that he was putting England on notice that vaccine passports would be *required* in certain settings, such as nightclubs and other large venues, from September, by which time all adults over the age of 18 would had been offered a second jab. What is more, proof of a negative test would no longer be enough.[110] He added that the government 'reserve[d] the right to mandate certification at any point if it is necessary to reduce transmission'.

'Why the fuck do we have to announce it now?' one senior Conservative MP said of this latest U-turn.[111] It was a fair question, although a better one would have been: 'Why do we have to *build* it now?' The United Kingdom was, after all, already doing well with its vaccination rollout, unlike countries such as France and Italy, where vaccine uptake was lower and slower until their governments introduced vaccine passports for domestic use to pressure people into getting vaccinated.

Dominic Raab, then the Foreign Secretary, told *The Times* that this was precisely the reason the British government was considering it. 'It is a little bit of coaxing and cajoling,' he said, while admitting that once we had pushed up our vaccination rate, 'the wider questions about vaccine certification become much less relevant and less salient'.[112] Alas, he never specified the government's target for vaccination, so we had no idea of what the government's idea of success looked like or how we would measure it – and thus we had no way of knowing if vaccine passports for domestic use would help us reach our goal or when we might be able to do away with them.

In the absence of such criteria, the debate continued. By September the prime minister had changed his mind yet again, announcing that the government would *not* require vaccine passports for domestic use in England.[113] Yet again he hedged his bets, though, saying that the government would keep the idea in reserve in case the virus overwhelmed the NHS in the winter of 2021–22, and that it might introduce this requirement with just one week's notice.

Health Secretary Sajid Javid flirted with setting out some criteria when he told BBC Radio 4's *Today* programme that 'there won't be any single trigger' but rather 'a number of measures we are going to keep under close watch with our friends in the NHS. That of course includes hospitalisations, it includes the pressure on A&E [accident and emergency], on the ambulance services, staffing levels.'[114] No further precision was given.

In October 2021 Scotland and Wales voted to require Covid vaccine passports for domestic use in certain settings.[115] Northern Ireland followed suit in November and England in December.[116]

As of this writing, the NHS Covid Pass allows us to show that we have had either a vaccine or a recent negative test. This means that, strictly speaking, it is not a vaccine passport for domestic use. However, this criterion can and will change.

The UK Health Secretary Sajid Javid has already told the House of Commons that 'once all adults have had a reasonable chance to get their booster jab', three doses of a vaccine will be required for full vaccination status. This will likely shift again if further boosters are required in 2022 and beyond.

Finally, while the NHS Covid Pass is currently only required for nightclubs and large events, both indoor and outdoor, the government could easily expand this to include pubs, restaurants, cinemas, theatres, gyms, hotels, etc. – as is already the case in many other countries.

All of these decisions come with trade-offs, because vaccines and vaccines passes/passports are not neutral tools and technologies. Where we draw the line on them in the United Kingdom has already changed dramatically in just a year. Regardless of where we personally draw the line, we are all technology ethicists now.

CONCLUSION

As we prepare for the coronavirus public inquiry in 2022, we will need to assess our track record in this pandemic so that we can compare the time and resources we have spent on each mitigation against our deaths and hospitalizations, against the number of cases of long Covid that we have, against other health, social and economic impacts, and against the strain that has been put on the NHS and on the social care system. We owe it to everyone we lost, to everyone who risked their lives to care for us and keep our society functioning, and to everyone who has worked so hard to get us out of this crisis to ensure that we learn the right lessons and emerge better prepared for future pandemics. We must determine how effective our new pandemic digital health tools were and how they compare with traditional disease control measures.

Our answers will likely vary according to the different stages of the pandemic. Before we began mass vaccination in December 2020, we had very few mitigations that demonstrated *a causal relationship* with lower case numbers, hospitalizations and deaths: lockdowns, working from home where possible, social distancing, mask wearing, hand hygiene and ventilation. During this stage, our priority was to stop as many infections as we could. QR code scanning and exposure notification apps may have helped with that, even with their limitations and the problems we experienced during their roll-out.

However, we must bear in mind that even when the tools themselves worked according to their design specification, the 'test–trace–isolate' system of which they were a part did not. That is neither the fault nor the responsibility of the technology or the technologists. Rather, it reflects the government's struggle to build a testing capacity that could deliver accurate, reliable test results within twenty-four hours and its failure to support many people so that they could afford to self-isolate when exposed to someone with a positive test for the virus.

QR code scanning and exposure notification apps can do little when set against such a broken system. Moreover, by being available only to people who had compatible smartphones, these technologies amplified healthcare inequalities across the population. They became redundant when enough of the population had been vaccinated or had survived infection and acquired temporary immunity, thereby reducing deaths and hospitalizations to a level the government deemed acceptable, which led to its decision to open up the country on 19 July 2021.

From that point on, the government no longer prioritized breaking the chain of transmission and keeping case numbers down. Rather, its focus was on limiting the number of hospitalizations, as these can overwhelm the NHS and stop it from being able to deliver care. Its emphasis on mass vaccination – which expanded from August 2021 to include children age 16–17 and then those aged 12–15 – shifted the focus on digital health tools from QR code scanning and exposure notification apps to vaccine passports for domestic use. That tool has a different purpose: to 'nudge' us to get vaccinated and boosted.

Prime Minister Boris Johnson has therefore presided over the creation of the very thing he once opposed: a national identity card, and one that is more intrusive than the one he rejected in the mid 2000s since it includes facial verification technology and health data. As of this writing, he has yet to eat it, masticate it or cut it up with a pen knife. The governments of the devolved nations of the United Kingdom have limited the use of the NHS Covid Pass, but they could change the law at any time and at short notice. They have given no indication of whether the NHS Covid Pass will ever be decommissioned or repurposed (e.g. to include vaccines against flu or other illnesses). Nor have they indicated what would happen to the NHS Covid Pass if a variant emerges against which vaccines are ineffective.

All we can say for certain is that it will take much more than an app – or indeed any single tool or technology – to solve a problem like a pandemic.

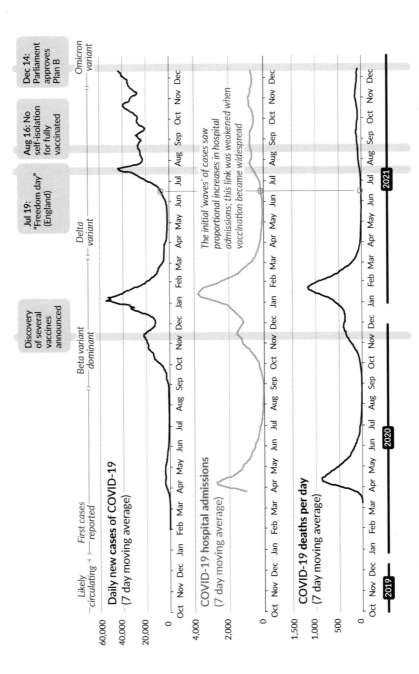

Figure 21. Timeline of the pandemic-fighting measures either considered or deployed in England (2020–2021).

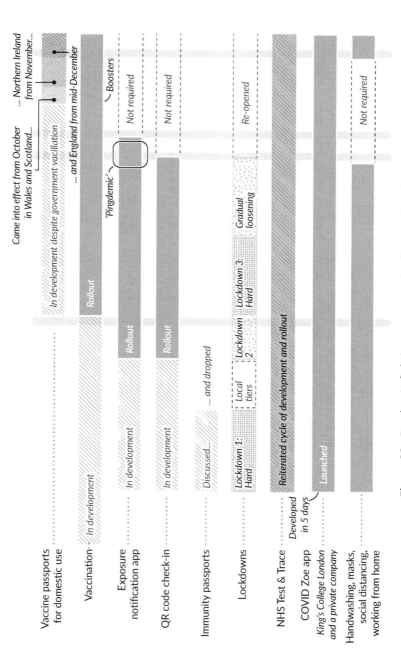

Figure 22. Pandemic-fighting measures timeline continued.

Conclusion

In October 2018 I was on a flight from London to Edinburgh to attend a conference where I was to present an early stage of the research that would become this book. The weather was so terrible that most flights had been cancelled, and our plane bumped along uncomfortably from turbulence. As we began our descent, we struggled against the strong wind. We dipped left, then right. I gulped as I looked through the window and saw white caps cresting on the choppy blue waters below. Down, down, down the pilot brought the plane. An uneasy silence came over the cabin as we braced for what promised to be a rough landing, only for a collective gasp to erupt as we suddenly took off with a shocking thrust of power. As we climbed high and fast, the smell of someone being sick wafted through the air along with the sounds of people crying. The passenger sitting next to me, a complete stranger until this moment, reached out to hold my hand. I squeezed back. Once we were steady and cruising, the pilot made an announcement over the loudspeaker: we were heading back to London. Amid groans of disbelief he explained that the plane's computers had decided that it was unsafe to attempt the landing, and he agreed. He did not want to risk the lives of everyone on board, and he hoped we would understand.

As we turned around, I wondered what would have happened if the pilot had *disagreed* with the computer. Would he have overridden it in favour of his own judgement and attempted the landing? What would have happened in terms of how that airline's computers were programmed afterwards if he had been right – or wrong? Was that learning captured? I wanted to ask him these questions, but by the time we finally got back on the ground, all I could manage was to thank him.

For a long time I remained troubled by the question of what happens when a human and a computer disagree about what to do, especially when lives hang in the balance. It reminded me of an incident recounted by Taylor Downing in his nail-biting book *1983: The World at the Brink*, which I shall summarize below.

On 26 September 1983 Lieutenant-Colonel Stanislav Petrov was on duty at a military compound outside Moscow when the computer system began to sound the alarm for a missile attack.[1] As deputy chief of the Department of Military Algorithms, he had helped both to write the operating instructions for the software and to install it, and thus 'did not have the confidence that some of the other officers had in it'. Petrov figured it was a false alarm and, in the tradition of people everywhere who are dealing with suspected machine malfunction, told his team to turn the equipment off and then back on again.

Again the system flashed the launch message, and again Petrov thought it was a false alarm and ordered a reboot. Now it indicated that at least three missiles had been detected. 'Terrified, hot and sweaty', Petrov checked with the satellite analysts, who said they could see no sign of a launch, and with the computer analysts, who said there was no sign of a system malfunction. Who was right? Was this a false alarm or had the Americans launched a nuclear attack? Should the Soviets retaliate? Even as the system detected two more launches, Petrov drew on his knowledge and experience and maintained that it was a false alarm. The other senior officers present gradually realized that none of the other tracking stations had detected any missiles – the satellite and computer systems had failed.

For years most of the world remained ignorant of how close we had come to a nuclear war. As for Petrov, he was neither thanked nor honoured for his actions. On the contrary, he was reprimanded for failing to follow protocols, discharged from the military and later had a nervous breakdown. In the meantime, a generation of children continued to grow up, play, and go to school. Among them was a little girl living 8,000 kilometres west of Moscow. One day she would grow up and cite Petrov's example in her book about technology ethics, wishing she could thank him.

Two examples of human–computer decision-making, two different outcomes. In the first, lives were preserved because the human *agreed* with the computer; in the second, lives were spared because the human *disagreed* with the computer. This points to a problem that extends well beyond the 'trolley problem' most frequently applied to self-driving cars and who they should save or kill. Do we always want to have a human 'in the loop' when it comes to computer-assisted decision-making? Are there any areas in which we would be comfortable with allowing a machine, on its own, to decide our fate? If so, who is responsible – and can therefore be held accountable or liable – for the outcome? Where do we draw the line?

TOWARDS A CULTURE OF TECHNOLOGY ETHICS

The University of Illinois at Urbana-Champaign has long offered some of the finest computer science and engineering education in the United States. Alas, this was entirely wasted on me when I did my undergraduate studies there in the second half of the 1990s. As a French major who flirted heavily with other languages, history and literature, my interests lay elsewhere, literally: West Green Street divided the humanities quad from the engineering quad. I only ever crossed it to hang out with my friends, who were all studying computer science and various engineering disciplines, and I only took Computer Science 101 because it was a graduation requirement.

I do not recall any mention of technology ethics in that class, or in conversations with my friends, or in the general excitement of the dot.com era. It is not that the concept did not exist. As this book has noted, technology ethics and related disciplines such as science and technology studies, digital humanities and human–computer interaction have long flourished. Rather, it is as though there was a mental divide that was every bit as real as West Green Street – computer science and engineering did not cross over to hang out with the arts and humanities, and vice versa.

Thankfully, a shift is underway in how we are teaching the next generation. Technology ethics is increasingly on the curriculum, as the following examples show.[2]

- In 2017 Cornell began offering a course on 'information ethics, law and policy'.
- In 2018 Harvard and MIT began offering a joint course on 'the ethics and governance of artificial intelligence', Stanford began offering a course on 'ethics, public policy, and technological change', and the University of Texas at Austin introduced a course on 'the ethical foundations of computer science'.
- In 2019 the Knight Foundation said it would give nearly $14 million to existing programmes at Indiana University, Stanford, the University of Texas at Austin, the University of Wisconsin-Madison and Yale; and $25 million to new ones at Carnegie Mellon, George Washington, NYU, the University of North Carolina at Chapel Hill and the University of Washington, all to study the impact of technology on democracy.

In the United Kingdom, UK Research and Innovation (a non-departmental public body of the UK government) announced in 2018 that it was investing £100 million in sixteen Centres for Doctoral Training to train 1,000 PhD students in aspects of AI, including transparency, accountability and responsibility. In 2019 the University of Oxford received a £150 million gift to build a centre that will host, among many other things, a new institute for ethics in AI and other emerging technologies. Such initiatives align neatly with the government's investment in AI. Since 2014 the United Kingdom has invested more than £2.3 billion into AI. In October 2021 the government announced 2,000 new elite AI scholarships for disadvantaged young people as well as doubling the number of Turing Institute research fellows to 400.

Of course, not everyone is teaching technology ethics, and even if they were, it is unlikely that it would be sufficient *on its own* to help us create and use technology in ways that maximize the benefits and minimize the harms. Still, we have to start somewhere. It is encouraging to see more and more people working to create a culture that considers ethics in the creation and use of tools and technology, and builds bridges between the liberal arts and humanities and science and technology. Surely this is an improvement on the culture described by the mathematician John von Neumann in his 1954 testimony to a secret hearing held by the Atomic Energy Commission:

> We were all little children with respect to the situation which had developed, namely, that we suddenly were dealing with something with which one could blow up the world. *None of us had been educated or conditioned to exist in this situation, and we had to make our rationalization and our code of conduct as we went along.*[3] [Emphasis added]

Strengthening our education and conditioning in technology ethics can empower all of us to be more than just a cog in someone else's machine. This is especially true for our young generations. They have grown up under the shadow of a possible nuclear war (as of 2020, nine countries were known to possess nuclear weapons, and there were around 14,000 of them in existence). They have also only known a world of the internet, social media and apps; post-9/11 security fears and surveillance capitalism; diminishing privacy and increasing misinformation. Today they are living through a pandemic and confronting climate change. All of this is shaping their values, including their views on whether technology is neutral and where we should draw the line. Like every generation before them, they will have to draw that line again and again.

I hope that this book may be of some help. We have expanded beyond asking *where* to draw the line to explore *how* to draw it and *how* to know if what we are doing is working. We have considered *who* draws the line and *who* decides when it has been crossed. We have explored what we mean by intelligence and decision-making so that we are able to examine the concept of responsibility – for humans *and* machines. We have taken a look at the emerging role of 'technology ethicist' and how some of the people doing technology ethics are helping to maximize the benefits and minimize the harms. With our in-depth analyses of facial recognition technology and pandemic digital health tools, we have demonstrated how we can apply a philosophical framework to technology ethics challenges to better understand the problems we are trying to solve and identify the best ways to solve them.

Nevertheless, this book does not pretend to offer an exhaustive or definitive account of technology ethics, though. It is a minimum viable product that we can deploy now and, if we wish, develop into something that is full-featured. What might that look like?

THE PROBLEM WITH PROBLEMS

It is easy to fall in love with a problem. It goes something like this.

We identify a problem. We describe it, arguing over terminology and definitions. We map out how it relates to other considerations (e.g. operational ones, strategic ones, degree of urgency, etc.). We contextualize it, situating it within time (history) and space (geography) and according to how others are approaching it (benchmarking). We do an opportunity–cost analysis to determine what it will cost to solve/not solve this problem. We evaluate whether this is really the most valuable problem to solve or if our resources would be better spent elsewhere. We come up with a number of solutions and pilot the most promising to identify if any might work. We seek out criticism – from within and externally – to identify our blind spots, find flaws in our efforts, iterate and refine.

It is *very* tempting to linger in this phase. It provides political cover, because we are *seen* to be doing something. It can be lucrative for anyone who solves problems for a living, allowing us to transform pilots into projects, and projects into long-term 'enterprise' partnerships. While the problem solvers look busy and make money, those who gain from problems staying unsolved remain in power and those who suffer from unsolved problems must look elsewhere or stay stuck.

To get to solutions, we must push through this phase. It is not enough to diagnose a problem (*'What is the problem we are trying to solve?'*). We also need treatment and prognosis. We need actionable insights, with differentiation between actions that are 'nice to do', actions that we 'need to do' and actions we should *not* do. We must be able to articulate a definition of success (*'What does success look like?'*). We must be able to measure our progress (*'What metrics will we use to measure our journey from problem scoping to solution?'*). While we might benefit from comparing ourselves against our definition of success and against the progress of others who face a similar problem (benchmarking), not all problems allow for such comparison – especially when we are working on something new. In such cases we will have to lead, which means we must risk getting it wrong as we try to get it right.

But what should we do when we encounter problems that cannot be solved?

The concept of a 'wicked problem' was coined in 1973 by Horst W. J. Rittel, a professor of science of design, and Melvin M. Webber, a professor of urban planning, and it was summarized by John C. Camillus in the *Harvard Business Review* in 2008 as follows:

Wickedness isn't a degree of difficulty. Wicked issues are different because traditional processes can't resolve them... A wicked problem has innumerable causes, is tough to describe, and doesn't have a right answer... Environmental degradation, terrorism, and poverty – these are classic examples of wicked problems. They're the opposite of hard but ordinary problems, which people can solve in a finite time period by applying standard techniques. Not only do conventional processes fail to tackle wicked problems, they may exacerbate situations by generating undesirable consequences.[4]

The concept of wicked problems is relevant to technology ethics because we often use tools and technology to solve social problems. These cannot be solved, Rittel and Webber argue, because 'social problems are never solved. At best they are only re-solved – over and over again.' Nor can we judge our efforts at re-solving wicked problems according to 'yes or no' or 'true or false'. Instead, we must make do with '"good or bad", or more likely, "better or worse" or "satisfying" or "good enough"'.[5]

The challenge of making tools and technology more beneficial and less harmful can *itself* be a wicked problem. Sometimes there are innumerable reasons why they are causing harm and it is not clear what, if anything, would fix them. On such occasions we too might be tempted to think that 'the only way to guarantee that such a powerful tool isn't abused and doesn't fall into the wrong hands is to never create it', as Apple explained on its website in 2016, when it refused a request from the US government to bypass security protections on the iPhone lock screen.[6] That might be true in some cases, but more often than not, 'there are opportunities and harms, and our job is to maximize opportunities and minimize harms', as Tracy Pizzo Frey, who sits on two ethics committees at Google Cloud as its managing director for Responsible AI, told Reuters in 2021.[7]

If only we had a checklist.

TECHNOLOGY ETHICS IN ACTION

When I first began writing this book before the pandemic I drew up the following list of actions that I had seen, heard of or read about to 'do' technology ethics, from small steps to the big picture.

- Make Big Tech companies pay more tax.
- Reform research and development tax credits to reflect that innovation comes not only from STEM subjects but from the social sciences, the liberal arts and the humanities too.[8]
- Impose moratoriums and bans (e.g. on facial recognition technology).
- Regulate the data tracking industry.[9]
- Create legal protections and financial incentives for technology, especially whistleblowing (as already exists for banking).
- Require an Ethics and Society Review for all AI research that seeks grant funding.[10]
- Expand the remit of universities' Institutional Review Boards to include the ethics of research *funding*, not just the ethics of the research itself.
- Create and implement ethics guidelines/principles/frameworks/toolkits throughout organizations.
- Convene an Ethics Committee/Advisory Board that meets frequently to assess benefits and harms.
- Appoint a Chief Ethics Officer/Chief Ethicist who reports directly to the CEO.
- Require external auditing for organizations' data, algorithms and methodologies so that we can know whether their products and services are safe to allow into the marketplace.
- Encourage investor activism in the sphere of data/AI/technology ethics, similar to that which exists in the environmental, social and governance (ESG) space.
- Update workers' rights to protect them from workplace surveillance.

- Foster the activism of technology workers (unionized versus non-unionized).
- Conduct design thinking/consequence scanning/due diligence/ethics reviews.
- Teach technology ethics to journalists, the police, the military, regulators and lawmakers.
- Improve digital literacy by including age-appropriate technology ethics in the education curriculum from kindergarten to tertiary level.
- Create a digital Bill of Rights to protect people in terms of data protection, privacy, civil liberties, human rights and being on the receiving end of AI.

I planned to evaluate each of the above items against a set of criteria or metrics so that we could determine which ones were essential, optional or not worth doing. For example:

- *What* action should be taken?
- *Why* take this action?
- *Who* should take this action?
- *What* does success look like?
- *How* we will know when we have reached it?
- *Who* is already doing this?
- *Which experts* can advise and help test this out?
- *Who/what* is in scope and out of scope? Who is this action for, and who are we ignoring or omitting?
- *Impact:* on whom will our tool or technology be used, and who will be harmed by it and who will not be?
- *How* will we know if we are wrong?
- *Verdict:* is this action worth taking? (Consider cost, ease of doing, is it a voluntary step or a legal requirement, etc.)

Then the pandemic hit and many things took a different direction – my thinking included. While I would still be curious to see such an analysis, I now think technology ethics is often a wicked problem and thus requires an approach that is more suited to complex systems than linear problems. The pandemic is one example of a complex system, but so are climate change, cybersecurity failures, supply chain dysfunction, biodiversity loss, inequality and the energy transition, to name just a few. No one person, organization or country can solve them alone; we will have to work together. No single mitigation is likely to be a silver bullet; a combination of mitigations, continually refined, may prove more effective.

So where does that leave us in terms of actions we can take as individuals?

DO WE NEED A HIPPOCRATIC OATH FOR TECHNOLOGY?

In *Hello World: How to Be Human in the Age of AI*, Hannah Fry, a professor of mathematics at University College London, called for mathematicians, computer engineers and scientists in related fields to take a Hippocratic Oath, observing that:

> In medicine, you learn about ethics from day one. In mathematics, it's a bolt-on at best. It has to be there from day one and at the forefront of your mind in every step you take...
>
> We've got all these tech companies filled with very young, very inexperienced, often white boys who have lived in maths departments and computer science departments. They have never been asked to think about ethics, they have never been asked to consider how other people's perspectives of life might be different to theirs, and ultimately these are the people who are designing the future for all of us.[11]

Fry's thinking was similar to that of Sir Joseph Rotblat, who had called for 'A Hippocratic Oath for Scientists' in 1999.[12] His appeal carried weight for two reasons. First, Rotblat was the only scientist to quit the Manhattan Project on the grounds of conscience. Second, it came four years after Pugwash – the group Rotblat cofounded in 1957 with Einstein and the philosopher Bertrand Russell to educate the world about nuclear weapons and weapons of mass destruction – had won the Nobel Peace Prize. Rotblat endorsed the following text, which was created by Student Pugwash USA and could serve as a Hippocratic Oath for science and technology:

> I promise to work for a better world, where science and technology are used in socially responsible ways. I will not use my education for any purpose intended to harm human beings or the environment. Throughout my career, I will consider the ethical implications of my work before I take action. While the demands placed upon me may be great, I sign this declaration because I recognize that individual responsibility is the first step on the path to peace.

Rotblat, in turn, echoed Karl Popper, an Austrian-born British philosopher who also thought that scientists had a moral responsibility – but with a twist. When he gave his public address in 1968 on the 'The moral responsibility of the scientist', he included social scientists too:

> The social scientist has a particular responsibility here, because his studies more often than not concern the use and misuse of power pure and simple. I feel that one of the moral obligations of the social scientist which ought to be recognised is that, if he discovers tools of power, especially tools which may one day endanger freedom, he should not only warn the people of the dangers but devote himself to the discovery of effective counter-measures...
>
> The problem of the unintended consequences of our actions, consequences which are not only unintended but often very difficult to foresee, is the fundamental problem of the social scientist.[13]

As we think about how to apply the Hippocratic Oath to technology, we might consider how it works in medicine. Not all medical schools today invite their students to take a Hippocratic Oath, but those that do can choose from a number of modern versions or even create their own with their students.[14] Either way, a modern Hippocratic Oath refers to the four principles of medical ethics, which are

- *autonomy* (respect individuals' right to make their own choices and have their own life plan, which relates to the concept of *informed consent*),
- *non-maleficence* (do not inflict harm intentionally),
- *beneficence* (act for the benefit of others) and
- *justice* (distribute health resources fairly).[15]

This is not to suggest that a list of principles or the taking of an oath will, on its own, minimize all harms. However, inviting ourselves to think about the values we want to guide the creation and use of tools and technology is one step that we can all take to create a culture of technology ethics. After all, in the critical moment when the creation or use of a tool or technology hangs in the balance, it is humans who must decide where to draw the line. It might be a good idea for them to have reflected on technology ethics beforehand rather than simply deciding in the moment.

The good news is that we do not need to be scientists, social scientists or technologists to do this. We all have a part to play in shaping our values and culture. While the challenges, risks and threats that face us are many and often daunting, we are not powerless against them. Businesses respond to customer demand. Elected officials respond to voters. Parents and teachers respond to the need to protect and nurture children. Children inspire adults to do better and often show us new ways to *be* better.

Ultimately, technology ethics is more than a philosophical framework that we can apply to technology to maximize its benefits and minimize its harms. It is an idea, and like stories or viruses, ideas have chains of transmission, spreading from one person to the next. I cannot wait to encounter yours.

Glossary

1:1 face matching   A form of facial recognition technology in which a 'live' image of us is matched to a record of us stored on a device or a database, e.g. when we unlock our smartphone or to access a government service, such as the NHS Covid Pass app, or use the biometric chip in our passport to clear customs and immigration at the airport.

1:many face matching   A form of facial recognition technology in which a 'live' image of us is checked against an entire database, e.g. when the police use live facial recognition technology (LFRT) in public to look for protestors, or when shops use LFRT to look for shoplifters.

Aadhaar   The world's largest biometrics project: a multimodal biometrics solution (iris, fingerprints, face) of more than a billion Indian citizens' and residents' data.

accessibility/accessible   The Web Accessibility Initiative defines this as 'addressing discriminatory aspects related to equivalent user experience for people with disabilities'. It goes on: 'Web accessibility means that people with disabilities can equally perceive, understand, navigate, and interact with websites and tools. It also means that they can contribute equally without barriers.'[1] *See also* 'inclusivity/inclusive'.

adoption rate   A metric used by app developers (and other product launchers) to measure how many people have downloaded their app, how many are using it and how often, etc.

aesthetics   Branch of philosophy concerned with questions of beauty and that which can be experienced. Key question in this book: what is experience?

affective technology   Also called emotion detection technology. Based on two highly contested premises: first, that all humans have a set of universal emotions; and second, that it is possible to read someone's emotions from their face. Yet as Kate Crawford notes: 'A comprehensive review of the available scientific literature on inferring emotions from facial movements published in 2019 was definitive: there is *no reliable evidence* that you can accurately predict someone's emotional state from their face.'

algorithm   A sequence of rules and instructions to solve a problem.

algorithmic transparency and accountability   The scrutiny of data-sets, code and algorithms to test for bias and accuracy and to see if their results can be reproduced (a key criterion of the scientific method).

anthropometry   Measuring humans.

anti-vaxxers   People who oppose vaccines.

applied ethics   Ethics that we put into practice.

asymptomatic   When we carry a disease or infection but experience no symptoms.

atavism   The belief that criminality is inherited.

augmented reality   Technology that 'takes computer-generated images and overlays them onto our view of the world' using the camera on our smart devices.

authorization   Being given the go ahead to do what you are allowed to do once you have proven who you are.

Belt and Road Initiative (BRI)   China's global infrastructure and invest-ment project launched in 2013 has four parts: a land component (e.g. railways), a maritime component (e.g. ports), a financial com-ponent (e.g. loans) and a digital component (whereby countries all over the world use Chinese technologies to monitor, analyse and control their populations).[2]

biometrics   A technology that turns your body into data and code. It includes our DNA, fingerprints, face, voice and other physical and behavioural characteristics.

black box   'It can refer to a recording device, like the data monitoring systems in planes, trains and cars. Or it can mean a system whose workings are mysterious; we can observe its inputs and outputs, but we cannot tell how one becomes the other.'[3]

centralized model   A database model that stores all the data in one database. The advantage is that it allows easy analysis of the data by whoever can use the database, which would have been very useful for epidemiologists, public health authorities and the gov-ernment during the pandemic. An app built on this model could have allowed for contact tracing rather than simply exposure notification. However, this database model offers less privacy for individuals and poses a greater cybersecurity risk. *See also* decentralised model.

chain of transmission   How a virus spreads from one person to another.

citizen science   Scientific research conducted entirely or in part by amateur/non-professional scientists, including ordinary people and children.

cognition   The process of knowing.

consciousness   To have both self-awareness and perception of our surroundings, as well as an awareness of our eventual death.

contact tracing   An activity done by human contact tracers, who are usually public health officials. They work with an infected person to map out everyone they have come into contact with during the period of virus transmission. These people are then contacted and *their* contacts are mapped out, and so on. The aim is to map out the chain of transmission and then break it.

Coronavirus Act 2020   Legislation enacted in the United Kingdom in 2020 setting out new rules and regulations with respect to the pandemic.

Covid-19   The disease that sometimes results when people are infected with the virus called severe acute respiratory syndrome coronavirus 2 (SARS-CoV-2) and its variants.

Covid long-haulers   People who have been infected with the novel coronavirus, experience symptoms and take longer than a few weeks to recover. Some take weeks and months to recover; as of this writing, some are still experiencing symptoms after a year.

cryptography   Code-making and code-breaking.

cybernetics   The theory of communication and control processes in animals and machines.

cybersecurity   Measures taken to protect data, devices and technical infrastructure against unauthorized access.

data visualization   Techniques to communicate data and information. The use of data visualization to influence public health policy was pioneered by Florence Nightingale, a nurse and the first female fellow of what became the Royal Statistical Society.

debug   To find errors in code and then correct them.

decentralized model   A database model that does not store the data in one database. Rather, storage is distributed: in other databases or on people's devices (this is known as 'edge computing'). The advantage is that it offers greater privacy for individuals and lower cybersecurity risk. The disadvantage is that it provides less data for epidemiologists, public health authorities and the government during a pandemic. *See also* centralized model.

deductive arguments   A form of logic that starts with general conditions and moves to specific conclusions.

deep learning   A subset of machine learning based on artificial neural networks.

digilantism   Online vigilantism.

digital humanities   Applies computer-based technology to the humanities to create digital tools, methodologies, archives and databases researching and analysing content and data.

digitally excluded   People who cannot access tools or technology because they do not have the financial means, or because their existing tools and technologies are not compatible with a new offering.

disinformation   False information deliberately and often covertly spread in order to influence public opinion or obscure the truth.

drone   An unmanned aerial vehicle.

dual-use technology   Can be used for either civilian or military purposes.

Electronic Numerical Integrator and Computer (ENIAC)   A digital computer used by the US Army to calculate ballistics trajectories during World War II. It was programmed by six women: Kay Mauchley Antonelli, Jean Bartik, Betty Holberton, Marlyn Meltzer, Frances Spence and Ruth Teitelbaum.

embodiment   Our existence within our body.

emotion detection technology   *See* affective technology.

encryption   The process of converting data or information into a code to prevent unauthorized access.

epistemic trespassing   When people use their expertise in one topic in order to claim expertise in other topics.

epistemology   Branch of philosophy concerned with knowledge, the methods for acquiring it, and its limitations. Key question for this book: how can we know?

ethics   Branch of philosophy concerned with right and wrong, how we should live, and what constitutes a good life, as well as where our values come from and on what basis we hold those values. Key question for this book: how should we live?

eugenics   The idea of using social control to improve the health and intelligence of future generations. This idea inspired many US states to introduce laws for the sterilization of criminals, the mentally ill and the intellectually disabled; and Nazi Germany to pursue a racial hygiene movement, which culminated in the extermination camps of the Holocaust.

existential risk   A risk to the very existence of humanity. Examples include a pandemic, climate change, alien invasion, a meteor hitting Earth and destroying it, a gigantic solar flare.

exposure notification app   Digital health tool developed by many countries during the pandemic to alert people when they came within two metres of someone who tested positive for the virus within the period of infection risk.

Exposure Notification Application Programming Interface (API)   A joint effort between Apple and Google in 2020 to provide the core functionality for building smartphone apps to notify users of possible exposure to confirmed Covid-19 cases.

facial analysis: classification   A form of facial recognition technology that can be used to analyse our characteristics in order to classify us according to criteria. Allows others to make decisions about us – sometimes with our knowledge and consent, often without. Classification of people based on characteristics may seem straightforward until we remember that ethnicity and race are constructed and highly contested categories. How we identify ourselves may not be the same as how we are identified by face technologies. This also applies to other ways people might choose to classify us, from our sexual orientation to our political orientation or even our emotions.

facial analysis: physical   A form of facial recognition technology that analyses our physical health via our face. Allows others to make decisions about us – sometimes with our knowledge and consent, often without.

facial recognition technology   A biometric technology that translates our face into data and computer code so that we can then be identified, verified and authenticated, both by ourselves and by others. Depending on who is using it and how, it can also be a surveillance technology.

friction   In technology design, friction is sometimes deliberately included in a process in order to slow it down or make users think (e.g. Twitter asking its users if they would like to read an article before they retweet it). Other times, friction is the result of poor design and unnecessarily slows down a system.

Global Partnership on Artificial Intelligence (GPAI)   Created in 2020 by fifteen founder members (Australia, Canada, France, Germany, India, Italy, Japan, Mexico, New Zealand, the Republic of Korea, Singapore, Slovenia, the United Kingdom, the United States and the European Union) to guide the responsible development and use of AI.

herd immunity   Resistance to the spread of an infectious disease within a population based on immunity from either vaccination or previous infection.

holistic   In *philosophy*: the belief that the parts of something are intimately interconnected and explicable only by reference to the whole. In *medicine*: the treatment of the whole person, taking into account mental and social factors, rather than just the symptoms of a disease.

human centred design   A design philosophy that takes into account every person affected by the design of a tool, technology, product, service, process, etc.

human–computer interaction (HCI)   Focuses on the interfaces between humans (users) and computers, and thus sits at the intersection of several disciplines such as design, behavioural science, computer science and many more.

identity   Who you are as an individual and how you identify yourself in terms of characteristics and community.

immunity passport   This was an idea floated in the United Kingdom during the early months of the pandemic but not then tried. It would have allowed people who had recovered from being infected with the virus to resume public life while everyone else was still in lockdown, and it was abandoned after we realized it would encourage some people to get infected.

inclusivity   The Web Accessibility Initiative defines this as being about 'diversity, and ensuring involvement of everyone to the greatest extent possible. In some regions this is also referred to as *universal design* and *design for all*. It addresses a broad range of issues including accessibility for people with disabilities; access to and quality of hardware, software and internet connectivity; computer literacy and skills; economic situation; education; geographic location; culture; age, including older and younger people; and language.'[4] *See also* accessibility/accessible.

inductive arguments   A form of logic that starts from specific conditions and moves to a general conclusion.

infodemic An overabundance of information and the rapid spread of misleading or fabricated news, images and videos.

interoperability The ability of software, systems and even organizations and individuals to work together seamlessly.

'just in case' model A way of running some branches of the government, the economy and individual businesses before the pandemic. It encouraged stockpiling and strategic planning in order to better cope with supply chain risk or other risks. Examples include pandemic preparedness (although most countries failed at that), a strategic oil reserve, cyber resilience and business continuity planning.

'just in time' model A way of running many branches of government, the economy and individual businesses before the pandemic. It discouraged stockpiling and encouraged 'lean' operations and 'just in time' delivery. The goal here is to maximize profits and shareholder value. Reserves are seen as wasteful and unprofitable on the balance sheet.

LiDAR (light detection and arranging) A remote sensing technology that uses pulsed laser energy (light) to measure distances. It is used in some smartphone cameras, self-driving cars, robotics, drones and VR headsets.

lockdown This can be severe, in which a country shuts down public life and requires its inhabitants to stay home, or limited, in which a country restricts only some parts of public life. Lockdowns can also be localized within a country in order to deal with localized outbreaks.

logic A branch of philosophy and a subfield of mathematics and language; a systematic way of working through the implications of true statements; we use it to determine whether an argument is sound or if a hypothesis is false. Key question for this book: how do we know what we know?

machine learning A subfield of artificial intelligence. 'Statistics on steroids', as Meredith Broussard described it. At present, most artificial intelligence is machine learning. Deep learning is a subset of machine learning.

map A representation of reality, such as a map of the world; a representation of a concept, such as the London Tube map; a map of relationships; a map of connections between ideas and concepts; Alexander von Humboldt's 'web of life'.

metadata   Data about data; includes all the information associated with every email, text, social media post, digital photo and video we send.

metaphysics   Branch of philosophy concerned with the study of the nature, structure and origins of the universe – in other words, reality. Key question in this book: what is reality?

metoposcopy   The belief that a person's character can be gleaned from their physical appearance, such as their facial expressions or simply their forehead wrinkles.

minimum viable product   A version of a product with just enough features to be usable by early users who can then provide feedback for future development.

misinformation   Incorrect or misleading information.

mission creep   A gradual shift in objectives, often resulting in an unplanned long-term commitment – and unintended consequences.

National Institute of Standards and Technology (NIST)   A US federal laboratory that develops standards for new technology and is part of the US Department of Commerce.

neural networks   A means of doing machine learning whereby a computer learns to perform some task by analysing training examples that are usually labelled in advance. Neural networks are modelled loosely on the brain and must be trained.

neuro-rights   A concept that some people have proposed should be added to the Universal Declaration of Human Rights since emerging neurotechnologies may alter what makes us human.

niche   Niche tools or technologies are used by fewer people, usually because they are more expensive, offering the possibility of something bespoke, tailored or luxury. They are the opposite of tools or technologies that scale (which means that they can be used by the most people).

panopticon   A prison in which the warden can see any inmate at any time but the inmates can never be certain whether they are being watched – and so logically should assume that they are being watched at all times. An idea of Jeremy Bentham, who expanded upon an idea from his brother Samuel Bentham for worker surveillance. Expanded upon further by Michel Foucault.

personal protective equipment (PPE)   Equipment worn by health professionals, care workers and some other key workers during the pandemic to protect them from the virus. Includes masks, visors, gloves and, in some cases, full body gear.

perverse incentives Incentives that create unintended negative consequences. For example, immunity passports (which were considered before vaccines were developed) encouraged people to get infected in order acquire immunity (assuming that they survived!) so that they could regain their freedoms during lockdown – action that would have resulted in higher transmission rates.

phrenology Debunked pseudoscience that argues that an individual's abilities and criminality could be determined by studying the contours of the head.

physiognomy Debunked pseudoscience that argues that a person's character can be gleaned from their physical appearance – usually their facial expressions.

political philosophy Branch of philosophy that explores the relationship between the individual and the state. It examines concepts such as power, authority, legitimacy and freedom. Key question for this book: what is the nature of power?

predictive policing There are two broad types of predictive policing algorithms: location based and person based. The first use tries to figure out where a crime will happen; the second tries to figure out who is likely to commit a crime or reoffend.

pre-symptomatic The status of a person who is infected with a virus but has yet to show symptoms.

programmed behaviour Examples are a hermit crab 'knowing' to carry the shell of another creature, squirrels 'knowing' to store food, and animals that use intelligence to use or make tools, such as chimpanzees making spears for hunting or wild dolphins using a sponge to flush out prey.

proof of concept A very rough example or mock-up of an idea, tool or technology. This comes at the very beginning of turning an idea into reality, and it cannot be 'launched' to others. It represents an earlier stage than 'minimum viable concept', which can be launched.

realpolitik A system of politics or principles based on practical rather than moral or ideological considerations.

recidivism The tendency of a convicted criminal to reoffend.

resilience Ability to withstand stress, strain and shocks over the short, medium and long term.

right to an explanation   The right to information about individual decisions made by algorithms, e.g. decisions about credit, employment, etc. The right is guaranteed by Article 22 of the EU's data protection laws (which the post-Brexit United Kingdom is looking to remove).[5]

robot   A machine capable of carrying out a complex series of actions automatically.

scale   Refers to whether a tool or technology can be used by the most people. It is the opposite of *niche* tools or technologies, which are used by fewer people, usually because they are more expensive, offering the possibility of something bespoke, tailored or luxury.

Science and Technology Studies (STS)   Explores the relationship between science and technology, and society, politics and culture.

scientific method   Scientists use the scientific method. They make observations, develop a hypothesis to explain what they have observed, test the hypothesis with experiments, and then publish the results so that other scientists can see if they can reproduce them. If they cannot reproduce the results – and even when they can – they offer critiques and suggestions for what to do next. Yet the scientific method emerged from a much older system of thought – one that is well suited to the question of where we should draw the line: philosophy.

sentience   The ability to feel, perceive and experience, with a particular emphasis on touch.

signals intelligence   The branch of intelligence concerned with monitoring, intercepting and interpreting radio and radar signals.

singularity   The moment when machines surpass humans in intelligence.

social distancing   A behavioural regulation introduced during the pandemic instructing people to stay at least two metres apart from each other (in the United Kingdom; the distance varied in other countries). Also included rules about physical contact, including hugging and sex, as well as how close people could be in pubs, restaurants, in parliament and during protests.

software engineering   Branch of computer science that deals with the design, implementation and maintenance of complex computer systems.

superintelligence   An entity that surpasses humans in overall intelligence or in some particular measure of intelligence. In this book, it refers to fears that artificial intelligence will one day surpass human intelligence.

superspreading event An event in which an infectious disease is spread much more than usual. A person can also be a 'superspreader'.

symbolic tool use When a tool represents something else, such as money, or when it is used to change an emotional state, such as when we use a special blanket or doll to comfort a child.

technosolutionism The idea that technology is the solution to any problem. Exemplified by the phrase 'there's an app for that' when, in reality, apps are not a solution for most problems.

temperature checks Temperature checks have been used in Asia during the pandemic but criticized in the United States and the United Kingdom, where they were rarely used, because they are unreliable indicators of infection and could even encourage people to think they are not infected when they are and simply do not have a fever, thus allowing them to continue circulating and spreading the virus. *See also* thermal imaging.

test–trace–isolate This is the 'holy trinity' that public health authorities worldwide maintain is critical to suppress the virus. In the United Kingdom, we instead adopted an unholy pentagram of 'find–test–trace–isolate–support'.

transmission rate The number of other individuals that each infected individual will go on to infect in a population with no resistance to the disease.

transmission strategy This refers to all viruses, which must strike a balance as they spread: if they kills too many hosts too quickly, they will not be able to keep spreading and will burn out. The most successful viruses are those that can spread the furthest and last the longest, so they sometimes mutate into more contagious variants as they go. Their transmission strategy must allow enough hosts to live long enough to spread the virus.

trolley problem A thought experiment used to work out, for instance, how self-driving cars should 'decide' which humans to kill or spare. It was invented by two philosophers, Philippa Foot and Judith Jarvis Thompson, as a way of thinking through abortion and euthanasia.

Turing test A test devised by British computer scientist Alan Turing to determine whether a computer is capable of thinking like a human.

utilitarianism A philosophical concept pioneered by the British philosopher and jurist Jeremy Bentham (1748–1832), based on the principle of 'the greatest happiness of the greatest number'. This was expanded upon by the British philosopher and Member of Parliament John Stuart Mill (1806–1873).

vaccine hesitant   People who are reluctant to be vaccinated.

vaccine passport   Before Covid-19, vaccine passports were proof that we had received any vaccinations needed to travel internationally (e.g. for yellow fever), delivered in the form of a stamp in our passport. Since 2020, vaccine passports prove that we are fully vaccinated against Covid-19. These can be delivered digitally, in the form of a smartphone app, as well as physically, i.e. a paper printout with a QR code. Some countries have also deployed Covid vaccine passports for domestic use, e.g. to attend football matches or go to restaurants, the gym or the cinema.

value-sensitive design   Design methodology that asks who should be involved and what values are implicated in design in order to create tools and technologies that are more inclusive and accessible.

verification   How you prove your identity or credentials.

virtual reality (VR)   Technology that 'fully immerses people in 3D virtual environments' using headsets.

weapon of math destruction   Term coined by Cathy O'Neil in her book *Weapons of Math Destruction*. It refers to an algorithmic technology that can harm at mass scale.

wicked problem   A concept coined in 1973 by Horst W. J. Rittel, a professor of science of design, and Melvin M. Webber, a professor of urban planning, and summarized by John C. Camillus in *Harvard Business Review* in 2008 as follows: 'Wickedness isn't a degree of difficulty. Wicked issues are different because traditional processes can't resolve them... A wicked problem has innumerable causes, is tough to describe, and doesn't have a right answer... Environmental degradation, terrorism, and poverty – these are classic examples of wicked problems. They're the opposite of hard but ordinary problems, which people can solve in a finite time period by applying standard techniques. Not only do conventional processes fail to tackle wicked problems, they may exacerbate situations by generating undesirable consequences.'

Notes

Introduction

1   Peter Baker, Maggie Haberman and Annie Karni. 2021. Pence reached his limit with Trump. It wasn't pretty. *New York Times,* 12 January.
2   Twitter Inc. 2021. Permanent suspension of @realDonaldTrump. Blog post, 8 January.
3   Sara Fischer and Ashley Gold. 2021. All the platforms that have banned or restricted Trump so far. *Axios,* 9 January.
4   Joe Tidy. 2021. Silencing Trump: how 'big tech' is taking Trumpism offline. *BBC News,* 12 January.
5   James Clayton. 2021. Twitter boss: Trump ban is 'right' but 'dangerous'. *BBC News,* 14 January.
6   Alexander Mallin. 2021. At least 100 more to be charged in Capitol attack investigation, DOJ expects. *ABC News,* 12 March.
7   Kellen Browning and Taylor Lorenz. 2021. Pro-Trump mob livestreamed its rampage, and made money doing it. *New York Times,* 8 January.
8   David Yaffe-Bellany. 2021. The sedition hunters. *Bloomberg Businessweek,* 7 June. Amy Zegart. 2021. Spies like us: the promise and peril of crowdsourced intelligence. *Foreign Affairs,* July/August.
9   Eliot Higgins. 2021. *We Are Bellingcat: An Intelligence Agency for the People.* London: Bloomsbury.
10 Nicholas Fandos. 2021. Senate Republicans filibuster Jan. 6 inquiry bill, blocking an investigation. *New York Times,* 28 May.
11 Cecilia Kang. 2021. Lawmakers, taking aim at big tech, push sweeping overhaul of antitrust. *New York Times,* 14 June.
12 James Politi and Lauren Fedor. 2021. Lina Khan, the new antitrust chief taking on big tech. *Financial Times,* 18 June. Lauren Hirsh. 2021. Biden names a foe of Big Tech to head the Department of Justice's antitrust division. *New York Times,* 20 July.
13 The Editorial Board. 2020. Joe Biden, former vice president of the United States (interview). *New York Times,* Opinion, 17 January. Rachel Lerman. 2021. Social media liability law is likely to be reviewed under Biden. *Washington Post,* 18 January.
14 European Commission. 2021. The Digital Services Act package. Report (last updated 26 April).

15 Will Oremus and Aaron Schaffer. 2021. The Technology 202: China is doing what the US can't seem to do: regulate its tech giants. *Washington Post*, 28 July.

16 Cecilia Kang, David McCabe and Kenneth P. Vogel. 2021. Tech giants, fearful of proposals to curb them, blitz Washington with lobbying. *New York Times*, 22 June.

17 Nellie Bowles. 2019. Overlooked no more: Karen Spärck Jones, who established the basis for search engines. *New York Times*, 2 January. Karen Spärck Jones interview with Brian Runciman. 2009. British Computer Society, 3 May.

18 Ibid.

19 Martin Rees. 2018. *On the Future: Prospects for Humanity*, p. 165. Princeton University Press.

Chapter 1: Is technology neutral?

1 Paul Virilio. 1999. *Politics of the Very Worst: An Interview with Philippe Petit*, edited by Sylvère Lotringer, translated by Michael Cavaliere, p. 89. New York: Semiotext(e).

2 Joseph Weizenbaum. 1987. Not without us. *ETC: A Review of General Semantics* **44**(1, Spring), 44–45.

3 Tim Cook. Commencement speech to Stanford University. Cited in Lisa Eadicicco. 2019. Apple CEO Tim Cook called out companies like Facebook, Theranos, and YouTube in a speech pushing for responsibility in Silicon Valley. *Business Insider*, 17 June.

4 Jack Nicas and Katie Benner. 2020. FBI asks Apple to help unlock two iPhones. *New York Times*, 7 January. Jack Nicas, Raymond Zhong and Daisuke Wakabayashi. 2021. Censorship, surveillance and profits: a hard bargain for Apple in China. *New York Times*, 26 May. Joseph Cox. 2021. Apple delays release of child abuse scanning tech after backlash. *Vice*, 3 September.

5 Daniela L. Rus (director of the CSAIL of MIT). 2019. The future of AI. Video, Hub Culture Davos.

6 Paul Daugherty. 2019. A call for ethical leadership in technology and business. Speech to accept on behalf of Accenture the 'Corporate Honoree for Ethical Leadership' award from the Fellowships at Auschwitz for the Study of Professional Ethics (FASPE), 16 April.

7 Garry Kasparov. 2018. *Deep Thinking: Where Machine Intelligence Ends and Human Creativity Begins*. London: John Murray.

8 Garry Kasparov. 2019. AI ethics are human ethics. Tech is agnostic, it amplifies us. 'Ethical AI' is like 'ethical electricity'. Twitter, 16 January.

9 Jake Smeester. 2015. Agnostic technology: platform, protocol, and device. Blog post for ThingLogix, 4 August.

10 Dave Lee. 2019. Amazon's next big thing may redefine big. *BBC News*, 15 June.

11 Demis Hassibis. 2019. Interviewed by Jim Al-Khalili for the BBC. *Life Scientific*, 5 November.

12 A tweet by Adam Mosseri, head of Instagram (11 January 2021).

13 A tweet by Adam Mosseri, head of Instagram (11 August 2021).

14 Marc Andreesen was a guest on the podcast *The Rest Is History*, hosted by Tom Holland and Dominic Sandbrook. Silicon Valley Part 2 (7 September 2021).

15 Natasha Lomas. 2018. Tim Berners-Lee on the huge sociotechnical design challenge. *TechCrunch*, 24 October.

16 Steve Lohr. 2021. He created the web. Now he's out to remake the digital world. *New York Times*, 10 January.

17 John Thornill. 2020. World wide web founder scales up efforts to reshape the internet. *Financial Times*, 22 February.

18 Kate Crawford. 2018. Just an engineer: the politics of AI. Lecture at the Royal Society in London, 23 July.

19 Caroline Criado Perez. 2019. *Invisible Women: Exposing Data Bias in a World Designed for Men.* London: Penguin Random House.

20 Ibid., p. 163.

21 Ibid., p. 162.

22 Iris Bohnet. 2016. *What Works: Gender Equality by Design.* Cambridge, MA: Harvard University Press.

23 Sheila Jasanoff. 2016. *The Ethics of Invention*, pp. 5–6. New York: Norton.

24 James Titcomb. 2021. Facebook accused of discrimination after mechanic and pilot jobs targeted at men. *The Telegraph*, 8 September.

25 Ursula M. Franklin. 1990. *The Real World of Technology*, pp. viii, 2–3, 5. Toronto: Anansi (the revised 1999 edition has been used here).

26 Vannevar Bush. 1945. *Science: The Endless Frontier.* Report to President Harry S. Truman. Cited in Siddhartha Mukherjee. 2011. *The Emperor of All Maladies: A Biography of Cancer*, p. 120. London: HarperCollins.

27 I thank Professor Stuart Russell for adding some enhancements to this point. First, that 'the potential use of atomic energy for bombs was suggested by H. G. Wells in his novel *The World Set Free*, published in 1914, and explicitly discussed by Nobel laureate Frederick Soddy in 1915. Second, that Szilárd, who had read *The World Set Free*, had the original idea for neutron-induced nuclear fission in 1933 while watching the traffic lights turn green in Russell Square in London but did not then know which atoms would produce suitable neutrons when undergoing fission. Third, that Szilárd kept the 1934 patent on nuclear reactors secret and assigned it to the Admiralty, apparently to keep it from the Germans. And fourth, that in early May 1939 a team of nuclear physicists at the Collège de France in Paris led by Frédéric Joliot-Curie filed patent applications for nuclear technologies, including a nuclear reactor.

28 William Lanouette and Bella Silard. 1992. *Genius in the Shadows: A Biography of Leó Szilárd: The Man Behind The Bomb.* New York: Charles Scribner's Sons.

29 Ruth Lewin Sime. 1996. *Lise Meitner: A Life in Physics.* Berkeley, CA: University of California Press.

30 Lanouette and Silard (1992). *Genius in the Shadows.*

31 Emily Strasser. 2020. The bomb (a documentary). *BBC World Service*, 2 August.

32 Mark Coeckelbergh. 2020. *Introduction to the Philosophy of Technology*, pp. 5–7. Oxford University Press.

33 Gaia Vince. 2019. *Transcendence: How Humans Evolved through Fire, Language, Beauty and Time*, pp. 204–206. London: Penguin.

34 Andrea Wulf. 2016. The invention of nature. *Geographical*, 5 February.

35 Andrea Wulf. 2015. *The Invention of Nature: The Adventures of Alexander Von Humboldt, the Lost Hero of Science*, pp. 88–89. London: John Murray.

36   Corey Dickinson. 2021. Inside the 'Wikipedia of maps', a row over corporate influence. *Bloomberg*, 19 February.

37   Franklin (1990). *The Real World of Technology*, pp. viii, 2–3, 5.

38   Vince (2019). *Transcendence*, pp. 39–42.

39   Ibid., pp. 233–234.

40   Audrey Tang. 2019. A strong democracy is a digital democracy. *New York Times*, 15 October.

41   Kate Crawford and Vladan Joler. 2018. Anatomy of an AI system. Large-scale map and long-form essay. See https://anatomyof.ai/.

42   Interview with David Bowie by Jeremy Paxman. 1999. *BBC Newsnight*.

43   Marcel Duchamp (1887–1968) was a French artist and chess player who worked across several media and is associated with cubism, surrealism, Dada and conceptual art.

44   Coeckelbergh (2020). *Introduction to the Philosophy of Technology*, pp. 39–42. Marshal McLuhan. 1964. *Understanding Media: The Extensions of Man.* New York: McGraw-Hill.

45   Vince (2029). *Transcendence*, p. 217.

46   Ibid., pp. 220–221.

47   *BBC In Our Time*. 2019. Bergson and time. Presented by Melvyn Bragg with Keith Ansell-Pearson (Professor of Philosophy at the University of Warwick), Emily Thomas (Assistant Professor in Philosophy at Durham University) and Mark Sinclair (Reader in Philosophy at the University of Roehampton), 6 May.

48   Ibid.

49   Carlo Rovelli. 2019. *The Order of Time*, translated by Erica Segre and Simon Carnell. London: Penguin. Jim Al-Khalili. 2020. *The World According to Physics*, pp. 139–165. Princeton University Press.

50   Robert W. Shumaker, Kristina R. Walkup and Benjamin B. Beck. 2011. *Animal Tool Behaviour.* Baltimore, MD: Johns Hopkins University Press.

51   Dayglow Media & Pencil & Pepper. 2019. Why algorithms are called algorithms. *BBC Ideas* video, 9 July.

52   Shane Legg and Marcus Hutter. 2007. A collection of definitions of intelligence. In *Advances in Artificial General Intelligence: Concepts, Architectures and Algorithms (Proceedings of the AGI Workshop 2006)*, edited by B. Goertzel and P. Wang. Frontiers in Artificial Intelligence Applications, volume 157, pp. 17–24. IOS Press.

53   These questions on AI, embodiment and culture come from Alison Adam (professor of information systems at Salford University) in conversation with John Agar (lecturer in the history and philosophy of science at Cambridge University) and Igor Aleksander (professor of neural systems engineering at Imperial College London) on *In Our Time*, presented by Melvyn Bragg. *BBC Radio* (8 December 2005). See also 'Artificial intelligence', another episode of *In Our Time* presented by Melvyn Bragg, with Igor Aleksander and John Searle (professor of philosophy at the University of California). *BBC Radio* (29 April 1999).

54   Stephanie Hare chaired a discussion over Zoom with Martin Rees, Laura Mersini-Houghton, Hilary Lawson and Kate Devlin on the 'AI illusion' for the online festival 'How the Light Gets In' in May 2020.

55   'Artificial intelligence', episode of *In Our Time* (8 December 2005).

56   Max Tegmark. 2017. *Life 3.0: Being Human in the Age of Artificial Intelligence*, p. 49. London: Penguin.

57   Stuart Russell. 2019. *Human Compatible: AI and the Problem of Control*, pp. 13–61. London: Allen Lane.

58   Joanna Bryson. 2020. The artificial intelligence of the ethics of artificial intelligence: an introductory overview for law and regulation. In *The Oxford Handbook of Ethics of AI*, edited by Markus D. Dubber, Frank Pasquale and Sunit Das, p. 4. Oxford University Press.

59   Paco Calvo, Monica Gagliano, Gustavo M Souza and Anthony Trewavas. 2020. Plants are intelligent, here's how. *Annals of Botany* **125**(1), 11–28.

60   Zoe Schlanger. 2020. The botanist daring to ask: what if plants have personalities? *Bloomberg*, 21 November. Laura Spinney. 2020. Consciousness isn't just in the brain: the body shapes your sense of self. *New Scientist*, 24 June. Matthew Cobb. 2021. *The Idea of the Brain: A History*. London: Profile Books.

61   Vince (2019). *Transcendence*, p. 87.

62   René Descartes. 1637. *Discourse on the Method*.

63   Markus Weeks (ed.). 2019. *How Philosophy Works: The Concepts Visually Explained*, pp. 49–67, 146–147. London: Dorling Kindersley.

64   Coeckelbergh (2020). *Introduction to the Philosophy of Technology*, pp. 178–179.

65   Thomas Nagel. 1974. What is it like to be a bat? *Philosophical Review* **83**(4), 435–450. Coeckelbergh (2020). *Introduction to the Philosophy of Technology*, p. 436.

66   Alan Turing. 1950. Computing machinery and intelligence. *Mind: A Quarterly Review of Psychology and Philosophy* (October), 433–460.

67   Kenneth Cukier and Viktor Mayer-Schonberger. 2013. *Big Data: A Revolution that Will Transform How We Work, Live and Think*. Boston: Houghton Mifflin.

68   A tweet by Kenneth Cukier, senior editor at *The Economist* (24 October 2020).

69   Tom Simonite. 2021. This researcher says AI is neither artificial nor intelligent. *WIRED*, 26 April.

70   Jane Croft. 2021. AI system cannot be named as an inventor on a patent, UK court rules. *Financial Times*, 22 September.

71   I thank Stephanie Stoll and Ankur Banerjee for their critique of this illustration. For books I used Michael Wooldridge. 2018. *Artificial Intelligence*. London: Penguin Random House. Stuart Russell and Peter Norvig. 2016. *Artificial Intelligence: A Modern Approach*. Harlow: Pearson. Margaret Boden. 2018. *Artificial Intelligence: A Very Short Introduction*. Oxford University Press. Meredith Broussard. 2018. *Artificial Unintelligence: How Computers Misunderstand the World*. Cambridge, MA: MIT Press. Karen Hao. 2018. What is AI? We drew you a flowchart to work it out. *MIT Technology Review*, 20 November. Gary Marcus and Ernie Davis. 2019. *Rebooting AI*. New York: Vintage. Michael Wooldridge. 2020. *The Road to Conscious Machines*. London: Pelican. John Markoff. 2015. *Machines of Loving Grace: The Quest for Common Ground between Humans and Robots*. New York: Harper Collins. Kenneth Neil Cukier. 2019. Ready for robots? How to think about the future of AI. *Foreign Affairs*, July/August.

72   Broussard (2018). *Artificial Unintelligence*, pp. 10–11.

73   Nick Bostrom. 2014. *Superintelligence: Paths, Dangers, Strategies*. Oxford University Press.

74   Broussard (2018). *Artificial Unintelligence*, pp. 10–11, 32–33, 87–119.
75   Karen Hao. 2021. Stop talking about AI ethics. It's time to talk about power. *MIT Technology Review*, 23 April.
76   Will Knight. 2020. What AlphaGo can teach us about how people learn. *WIRED*, 23 December.
77   Russell (2019). *Human Compatible*, pp. 288–289.
78   Karen Hao. 2020. AI pioneer Geoff Hinton: 'Deep learning is going to be able to do everything'. *MIT Technology Review*, 3 November.
79   Ibid., pp. 290–291.
80   Frank Pasquale. 2015. *The Black Box Society: The Secret Algorithms that Control Money and Information*, p. 3. Cambridge, MA: Harvard University Press.
81   Cathy O'Neil. 2016. *Weapons of Math Destruction*. New York: Crown Publishing.
82   Broussard (2018). *Artificial Unintelligence*. Amy Thomson and Stephanie Bodoni. 2020. Google CEO [Sundar Pichai] thinks AI will be more profound change than fire. *Bloomberg*, 22 January. Rory Cellan-Jones. 2014. Stephen Hawking warns artificial intelligence could end mankind. *BBC News*, 2 December. Tom Simonite. 2021. This researcher says AI is neither artificial nor intelligent. *WIRED*, 26 April. Hao (2021). Stop talking about AI ethics. It's time to talk about power. Kate Crawford. 2021. *Atlas of AI: Power, Politics, and the Planetary Costs of Artificial Intelligence*. New Haven, CT/London: Yale University Press.

Chapter 2: Where do we draw the line?

1   Shane Richmond. 2010. How Google crossed the creepy line. *The Telegraph*, 25 October.
2   Avi Asher-Schapiro. 2020. Leading neuroscientists call for human rights principals to be extended to the brain. *Reuters*, 3 December.
3   Cited in Bruce Schneier. 2015. *Data and Goliath: The Hidden Battles to Collect Your Data and Control Your World*, p. 189. New York: W.W. Norton.
4   Schneier (2015). *Data and Goliath*, p. 20. John Naughton. 2016. Death by drone strike, dished out by algorithms. *The Guardian*, 21 February. Hayden's interview can be found in *The Johns Hopkins Foreign Affairs Symposium Presents: The Price of Privacy: Re-Evaluating the NSA* (7 April 2014).
5   Maria Giudice. 2019. Early Facebook designer: 'Sometimes you don't know something's wrong until it's too late'. *Fast Company*, 8 February.
6   Mukherjee (2010). *The Emperor of All Maladies*, p. 119.
7   Annie Jacobsen. 2014. *Operation Paperclip: The Secret Intelligence Program that Brought Nazi Scientists to America*. Boston, MA: Little Brown. Christopher Simpson. 1988. *Blowback: US Recruitment of Nazis and Its Effects on the Cold War*. New York: Grove. Tom Bower. 1987. *The Paperclip Conspiracy: The Hunt for the Nazi Scientists*. Boston, MA: Little Brown.
8   Michael Neufeld. 2017. The rise and fall of vengeance weapon 2. Video, Smithsonian National Air and Space Museum, 1 August.
9   E. H. S. Burhop. 1974. Scientists and soldiers. *Bulletin of the Atomic Scientists* (1 November), 271. Cited in Sarah Bridger. 2015. *Scientists at War: The Ethics of Cold War Weapons Research*. Cambridge, MA: Harvard University Press.

10 Julian Baggini. 2018. *How the World Thinks: A Global History of Philosophy*, p. xv. London: Granta. Citing Joram Tarusarira. 2017. African religion, climate change and knowledge systems. *Ecumenical Review* **69**(3), 408.

11 Gabriel Gatehouse. 2019. The puppet master (omnibus edition). *BBC*, 24 May. Peter Pomerantsev. 2019. *This Is Not Propaganda: Adventures in the War Against Reality*. London: Faber and Faber.

12 Lois Beckett. 2021. Facts won't fix this: experts on how to fight America's disinformation crisis. *The Guardian*, 1 January.

13 These questions are covered by a sub-branch of philosophy called philosophy of mind, as well as neuroscience. On philosophy of mind – which includes dualism, language, behaviourism, the mind–body identity theory, eliminative materialism, functionalism, biological naturalism and panpsychism – see Weeks (2019). *How Philosophy Works*.

14 Knight (2020). What AlphaGo can teach us about how people learn.

15 These definitions are from the Facebook Reality Labs.

16 Jordan Novet. 2021. Microsoft wins US Army contract for augmented reality headsets, worth up to $21.9 billion over 10 years. *CNBC*, 31 March.

17 Ibid.

18 Todd Haselton. 2019. How the Army plans to use Microsoft's high-tech HoloLens goggles on the battlefield. *CNBC*, 6 April.

19 William Gallagher. 2021. How to use the LiDAR scanner in iPhone 12 Pro. *Apple Insider*, 5 March.

20 Jeffrey Dastin. 2018. Amazon scraps secret AI recruiting tool that showed bias against women. *Reuters*, 11 October.

21 In the future it could consult the UK Information Commissioner's Office's 'Six things to consider when using algorithms for employment decisions'. Blog post, 18 December 2020.

22 Ben Weber. 2018. Governance in data science. *KDnuggets*, January. I thank David Matthew Millar for this point.

23 Melvyn Bragg and guests explore 'How can I know anything at all?'. *Omnibus: A History of Ideas*, BBC (7 April 2015). Miranda Fricker. 2007. *Epistemic Injustice: Power and the Ethics of Knowing*. Oxford University Press. Sun-Ha Hong. 2020. *Technologies of Speculation: The Limits of Knowledge in a Data-Driven Society*. New York University Press.

24 Fricker (2007). *Epistemic Injustice*.

25 Vidyasagar Machupalli. 2019. The hardest decision in the process of building a machine learning model is deciding on which algorithm to use. *IBM*, 13 August.

26 The Guardian view on DeepMind's brain: the shape of things to come. *The Guardian*, 6 December.

27 Michael Kearns and Aaron Roth. 2020. *The Ethical Algorithm: The Science of Socially Aware Algorithm Design*. Oxford University Press.

28 Peter Foster, Madhumita Murgia and Javier Espinoza. 2021. UK suggests removing EU's human review of AI decisions. *Financial Times*, 9 September.

29 Pasquale (2015). *The Black Box Society: The Secret Algorithms that Control Money and Information*.

30   Daisuke Wakabayashi and Scott Shane. 2018. Google will not renew Pentagon contract that upset employees. *New York Times*, 1 June. Hamza Shaban. 2018. Google employees go public to protest China search engine Dragonfly. *Washington Post*, 27 November.

31   Erin Griffith. 2018. Start-ups ask, 'Are we making money for Saudi Arabia?' *New York Times*, 1 November.

32   Wikipedia is 20, and its reputation has never been higher. *The Economist*, 9 January 2021.

33   *The Encyclopédie*. Victoria and Albert Museum website. Melvyn Bragg and guests Judith Hawley, Caroline Warman and David Wootton discuss *The Encyclopédie* on *In Our Time*, BBC (26 October 2006).

34   Oliver Balch. 2019. Making the edit: why we need more women in Wikipedia. *The Guardian*, 28 November.

35   Ibid.

36   Ibid. See also the work of Dr Jess Wade in the United Kingdom to increase the number of Wikipedia entries about women.

37   Merriam-Webster dictionary.

38   Joe Helm. 2021. 'Disinformation can be a very lucrative business, especially if you're good at it', media scholar says. *Washington Post*, 19 January.

39   Alisdair MacIntyre. 1977. Epistemological crises, dramatic narrative and the philosophy of science. *The Monist* **60**(4), 453–472. Philip Ball. 2020. Led by the science. *BBC*, 11 August.

40   Robinson Meyer. 2018. The grim conclusions of the largest-ever study of fake news. *The Atlantic*, 8 March.

41   Henry Mance. 2016. Britain has had enough of experts, says Gove. *Financial Times*, 3 June.

42   Nathan Ballantyne. 2019. Epistemic trespassing. *Mind* **128**(510), 367–395.

43   Elizabeth Dwoskin and Craig Timberg. 2021. Misinformation dropped dramatically the week after Twitter banned Trump. *Washington Post*, 16 January.

44   United Nations Department of Global Communications. 2020. UN tackles 'infodemic' of misinformation and cybercrime in COVID-19 crisis. Press Release, 31 March. World Health Organization. 2020. Immunizing the public against misinformation. Press Release, 25 August.

45   Philip Ball. 2020. The epidemiology of misinformation. *Prospect*, 19 May.

46   Nina Jankowicz. 2020. *How to Lose the Information War: Russia, Fake News, and the Future of Conflict*. New York: Bloomsbury. *The Economist*. 2021. A growing number of governments are spreading disinformation online (13 January).

47   Weeks (2019). *How Philosophy Works*.

48   Nor is this limited to philosophy. Economics and behavioural psychology are also interested in logic. See Peter Coy. 2020. Here are the 17 ways you make illogical decisions. *Bloomberg Businessweek*, 18 December.

49   Evan Andrews. 2018. 7 bizarre witch trial tests. *History Today*, 1 September. *History Today*. 2020. Salem witch trials. *History Today*, 23 October.

50   King James VI & I. 1597. *Daemonologie in Forme of a Dialogue, Divided into Three Bookes*. Cited in Sheilagh O'Brien. 2015. A 'divellish' woman discovered: the witchbury, 1643. *Cera: An Australian Journal of Medieval and Early Modern Studies* **2**, 10.

51 Wooldridge (2018). *Artificial Intelligence*, p. 6.

52 John Timmer. 2008. CAPTCHAs work? For digitizing old, damaged texts, manuscripts. *arsTechnica*, 14 August.

53 Ron Amadeo. 2017. Google's reCAPTCHA turns 'invisible', will separate bots from people without challenges. *arsTechnica*, 9 March. See also Google: 'What is reCAPTCHA?'

54 Josh Dzieza. 2019. Why CAPTCHAs have gotten so difficult. *The Verge*, 1 February.

55 Ibid.

56 Gus Hosein and Edgar Whitley. 2019. Identity and development: questioning Aadhaar's digital credentials. In Reetika Khera (ed.). 2019. *Dissent on Aadhaar: Big Data Meets Big Brother*, p. 222. Delhi: Orient Black Swan. Edgar A. Whitley and Gus Hosein. 2010. *Global Challenges for Identity Policy*. Basingstoke: Palgrave Macmillan.

57 Leo Kelion. 2021. Huawei patent mentions use of Uighur-spotting tech. *BBC News*, 13 January.

58 Kevin Roose. 2019. Why banning 8chan was so hard for Cloudflare: 'No one should have that power'. *New York Times*, 5 August.

59 Ibid. William Turton and Joshua Brustein. 2021. A 23-year-old coder kept QAnon online when no one else would. *Bloomberg*, 14 April.

60 Paul Karp. 2021. Google admits to running 'experiments' which remove some media sites from its search results. *The Guardian*, 21 January.

61 UK Information Commissioner's Office. 2018. Guidance to the General Data Protection Regulation. 25 May. BBC. 2020. EU–US privacy shield for data struck down by court. *BBC News*, 16 July. Nick Beckett. 2020. China passes draft law on personal data protection. *CMS-Law Now*, 22 October.

62 Baggini (2018). *How the World Thinks*, p. 294.

63 Terry Winograd and Fernando Flores. 1986. *Understanding Computers and Cognition: A New Foundation for Design*, p. xi. Boston, MA: Addison-Wesley.

64 Nelly Oudshoorn and Trevor Pinch (eds). 2003. *How Users Matter: The Co-Construction of Users and Technology*. Cambridge, MA: MIT Press.

65 Nir Eyal. 2014. *Hooked: How to Build Habit-Forming Products*. London: Penguin. Bill Burnett and Dave Evans. 2016. *Designing Your Life: How to Build a Well-Lived, Joyful Life*. London: Vintage.

66 UK Government Digital Services. 2019. Government design principles (10 September). Karwai Pun. 2016. Accessibility in government: dos and don'ts on designing for accessibility. UK Government, 2 September. Forbes. 2021. Deaf culture, deaf creatives and the quest for language exposure. *Forbes*, 4 February. Gareth Mitchell (presenter). 2021. Blindness in the digital age. *Digital Planet*, BBC (6 January).

67 Emma Parnell. 2021. Let's talk about sex*. Blog post, *Medium*, 19 March.

68 Haley Messenger. 2021. Blind Americans face roadblocks booking online vaccine appointments. *NBC News*, 13 March.

69 Ibid.

70 Andrew Leonard. 2020. How Taiwan's unlikely digital minister hacked the pandemic. *WIRED*, 23 July.

71 Sasha Costanza-Chock. 2020. *Design Justice: Community-Led Practices to Build the Worlds We Need*, p. 55–56, 223. Cambridge, MA: MIT Press.

72 Batya Friedman and David G. Hendry. 2019. *Value Sensitive Design.*
    Cambridge, MA: MIT Press.
73 Ed Catmull. 2014. *Creativity, Inc.* New York: Bantam Press. Steve Jobs. 2005.
    Commencement address delivered to Stanford University, 12 June. Niraj Choski.
    2016. The Trappist monk whose calligraphy inspired Steve Jobs – and influenced
    Apple's designs. *Washington Post,* 8 March.
74 David Kindy. 2019. How Susan Kare designed user-friendly icons for the first
    Macintosh. *Smithsonianmag.com,* 9 October. Tony Chambers. 2016. Sir Jony Ive
    reflects on the nature of objects, the fragility of ideas, and 20 years of Apple
    design. *Wallpaper,* 15 November.
75 Walter Isaacson. 2011. *Steve Jobs,* pp. 313–320, 41–45, 52, 525, 14, 31–33, 37. New
    York: Simon and Schuster. Shunryu Suzuki. 1970. *Zen Mind, Beginner's Mind.*
    Boulder, CO: Shambala. Jobs's appreciation for the aesthetic of Zen Buddhism
    worked well with Sir Jony Ive's admiration of Dieter Rams.
76 Hutcheson said, 'That action is best, which accomplishes the greatest
    happiness for the greatest numbers' in 1725. In the 1780s, Bentham proposed
    his 'felicific calculus'.
77 Andrew T. Forcehimes and Luke Semrau. 2019. *Thinking through Utilitarianism:
    A Guide to Contemporary Arguments.* Indianapolis, IN: Hackett.
78 Weeks (2019). *How Philosophy Works,* pp. 176, 180–181, 184–185.
79 Melvyn Bragg (presenter) and David Wooton, Helen Paul and John Callanan
    discuss 'The fable of the bees'. *In Our Time,* BBC (25 October 2018).
80 David Townson. Undated. The seven tenets of human-centred design. UK
    Design Council.
81 Eileen Guo. 2021. Tech is having a reckoning. Tech investors? Not so much.
    MIT Technology Review, 24 January.
82 Katherine Boyle. 2019. Goodbye, trolley problem. This is Silicon Valley's new
    ethics test. *Washington Post,* 5 February.
83 BBC. 2019. Drone no-fly zone to be widened after Gatwick chaos. *BBC News,*
    20 February. BBC. 2019. Changi Airport: drones disrupt flights in Singapore.
    BBC News, 25 June.
84 Gordon Corera (presenter) and Ben Crighton (producer). 2020. The new tech
    Cold War. *BBC Radio 4,* 23 June.
85 Nigel Inkster. 2020. *The Great Decoupling: China, America and the Struggle for
    Global Supremacy.* London: C. Hurst & Co.
86 Valentina Pop, Sha Hua and Daniel Michaels. 2021. From lightbulbs to 5G, China
    battles West for control of vital technology standards. *Wall Street Journal,*
    8 February. Anna Gross, Madhumita Murgia and Yuan Yang. 2019. Chinese tech
    groups shaping UN facial recognition standards. *Financial Times,* 1 December.
87 Yuan Yang and Madhumita Murgia. 2019. Facial recognition: how China
    cornered the surveillance market. *Financial Times,* 6 December.
88 James Kynge, Valerie Hopkins, Helen Warrell and Kathrin Hille. 2021.
    Exporting Chinese surveillance: the security risks of smart cities. *Financial Times,*
    9 June.
89 Shirley Zhao, Scott Moritz and Thomas Seal. 2021. Forget 5G, the US and China
    are already fighting for 6G dominance. *Bloomberg,* 8 February. Yang and Murgia
    (2019). Facial recognition.

90 World Bank. 2018. Belt and Road Initiative. Brief, 29 March. Lily Kuo and Niko Kommenda. 2018. What is China's Belt and Road Initiative? *The Guardian*, 30 July.

91 Paul Mozur, Jonah M. Kessel and Melissa Chan. 2019. Made in China, exported to the World: the surveillance state. *New York Times*. Nigel Inkster. 2021. *The Great Decoupling*, pp. 164–167. London: C. Hurst & Co.

92 Inkster (2019). *The Great Decoupling*, pp. 100–103.

93 BBC. 2021. Canada's parliament declares China's treatment of Uighurs 'genocide'. *BBC News*, 23 February. Reuters. 2021. Dutch parliament: China's treatment of Uighurs is genocide. *Reuters*, 25 February. Ben Mauk and Matt Huynh. 2021. Inside Xinjiang's prison state. *New Yorker*, 26 February. United Nations Human Rights Council, 46th session: Foreign Secretary's statement. UK Government, 22 February 2021.

Chapter 3: Facial recognition technology

1 Boris Johnson. 2004. Ask to see my ID card and I'll eat it. *The Telegraph*, 25 November.

2 Edgar A. Whitley and Gus Hosein. 2010. *Global Challenges for Identity Policies*. New York: Palgrave Macmillan.

3 Our *digital* national identity scheme is a work in progress: in 2021 the government scrapped the digital identity scheme for citizens to access government services it had been working on for eight years, and which cost the taxpayer £200 million, after several government departments refused to use it. Undeterred, it is trying again. Tony Diver. 2021. Ministers scrap digital ID project after spending eight years and almost £200m of taxpayers' money. *The Telegraph*, 19 March.

4 Ibid.

5 BBC. 2021. What should you do if stopped by police? *BBC News*, 1 October.

6 There are exceptions to this. Section 60 of the Criminal Justice and Public Order Act 1994 sets out the conditions in which the police can conduct a suspicionless search. The Coronavirus Act of 2020 has given the police expanded powers to stop and search people and to enforce lockdowns, which one serving London Metropolitan Police officer used to stop, kidnap, rape and murder Sarah Everard, a thirty-three-year-old woman who had been walking home during lockdown in March 2020. In response, the Met encouraged the public to challenge any lone police officer who stops us by asking, 'Where are your colleagues? Where have you come from? Why are you here? Exactly why are you stopping or talking to me?' If we still do not feel safe, we should consider 'shouting at a passerby, running into a house, knocking on a door, waving a bus down or, if [we] are in a position to do so, call 999', thereby summoning more police. What could go wrong? (See Legislation.gov.uk: Criminal Justice and Public Order Act 1994. See also Jamie Grierson. 2021. Sarah Everard case: people stopped by lone officer could 'wave down a bus' says Met. *The Guardian*, 1 October.)

7 Big Brother Watch. 2021. Britain's fight against ID: from war IDs to vaccine passports. Blog post, 21 February.

8   Ibid. See also politics.co.uk. Undated. Identity cards. Available at www.politics.
     co.uk/reference/identity-cards. The Guardian. 1952. Identity cards abolished
     after 12 years – archive. *The Guardian.*

9   BBC. 2020. Tony Blair: it is common sense to move toward digital IDs. *BBC News,*
     3 September.

10   Whitley and Hosein (2009). *Global Challenges for Identity Policies.*

11   The Identity Cards Act was given royal assent in 2006 and repealed in 2011.
     See UK Home Office. 2010. Identity cards are to be scrapped. 27 May. See
     also Alan Travis. 2010. ID cards scheme to be scrapped within 100 days. *The
     Guardian,* 27 May.

12   Theresa May. 2010. Home Office announces ID card scrap. 16 June.

13   Dame Cressida Dick, commissioner of the London Metropolitan Police. 2019.
     Speech to the Lowy Institute think tank in Sydney, Australia, 4 September.
     Martin Evans. 2019. Advances in technology risk turning society into a 'ghastly
     Orwellian police state', Met Commissioner warns. *The Telegraph,* 3 September.
     John Simpson. 2019. Beware Orwellian state, says Met chief Cressida Dick.
     The Times, 4 September.

14   Samuel Woodhams. 2021. London is buying heaps of facial recognition tech.
     WIRED, 27 September.

15   In 2021 the role of Biometrics Commissioner and Surveillance Camera
     Commissioner were merged. UK Parliament, House of Commons Science and
     Technology Committee. 2019. Issues with biometrics and forensics present
     significant risk to effective functioning of the criminal justice system. 18 July.

16   Mayor of London and London Assembly. 2019. Ethics panel sets out future
     framework for facial recognition software. 29 May.

17   UK Equality and Human Rights Commission. 2020. Facial recognition
     technology and predictive policing algorithms out-pacing the law. 12 March.
     Robert Booth. 2020. Halt public use of facial recognition tech, says equality
     watchdog. *The Guardian,* 12 March.

18   Jenny Rees. 2020. Facial recognition use by South Wales Police ruled unlawful.
     BBC News, 11 August.

19   Peter Fussey and Daragh Murray. 2020. Policing uses of live facial recognition
     technology in the United Kingdom. In *Regulating Biometrics: Global Approaches
     and Urgent Questions,* edited by Amba Kak, pp. 78, 81. AI Now Institute.

20   Spiegel International. 2012. German police identify burglar by his earprints.
     Spiegel International, 30 April.

21   Aaron Holmes. 2020. This smart toilet reads your anus like a fingerprint.
     Business Insider, 9 April.

22   Jane C. Hu. 2019. Say hello to dog facial recognition technology. *Slate,* 17 June.
     Sui-Lee Wee and Elsie Chen. 2019. China's tech firms are mapping pig faces.
     New York Times, 24 February.

23   BBC. 2018. Home secretary apologises for immigrant DNA tests. *BBC News,*
     25 October.

24   Helen Warrell. 2018. UK home secretary apologises to migrants forced to take
     DNA tests. *Financial Times,* 25 October.

25   Josh Taylor. 2019. Major breach found in biometrics system used by banks, UK
     police and defense firms. *The Guardian,* 14 August. Madhumita Murgia. 2019.
     The insidious threat of biometrics. *Financial Times,* 19 September.

26 Kevin Peachey. 2019. HMRC forced to delete voice files. *BBC News*, 3 May. Alan Travis. 2009. Home Office climbs down over keeping DNA records on innocent. *The Guardian,* 19 October. BBC. 2011. DNA and fingerprint guidelines 'unlawful'. *BBC News*, 18 May. Home Office. 2017. Custody images: review of their use and retention (24 February).

27 House of Commons. 2019. Facial recognition and the biometrics strategy, volume 659: debated on Wednesday 1 May 2019. UK Parliament, Hansard. See also UK Surveillance Camera Commissioner (who is being combined with the Biometrics Commissioner). 2020. Guidelines for police on use of LFR. Press release, 3 December.

28 Steven Feldstein. 2019. The global expansion of AI surveillance. Carnegie Endowment for International Peace, 17 September. Antoaneta Roussi. 2020. Resisting the rise of facial recognition. *Nature*, News Feature, 18 November.

29 Madhumita Murgia. 2019. Who's using your face? The ugly truth about facial recognition. *Financial Times,* 18 September.

30 Hannah Murphy. 2020. Meet the activists perfecting the craft of anti-surveillance. *Financial Times,* 25 June.

31 Todd Gustavson. 2009. *Camera: A History of Photography from Daguerreotype to Digital.* New York: Sterling.

32 Karin Andreasson. 2014. The first ever selfie, taken in 1839 – a picture from the past. *The Guardian,* 7 March.

33 Jane Caplan and John Torpey (eds). 2001. *Documenting Individual Identity: The Development of State Practices in the Modern World.* Princeton University Press.

34 Sir Arthur Conan Doyle. 1902. *The Hound of the Baskervilles,* p. 7. London: Penguin.

35 Leo Benedictus. 2006. A brief history of the passport. *The Guardian,* 17 November.

36 Jerone Andrews. 2021. The hidden fingerprint inside your photos. *BBC Future,* 25 March.

37 Sahil Chinoy. 2019. The racist history behind facial recognition. *New York Times,* 10 July. Crawford (2021). *Atlas of AI*, pp. 123–149.

38 Guillaume-Benjamin-Amand Duchenne de Boulogne. 1862. *Mécanisme de la physionomie humaine ou Analyse électro-psychologiques de l'expression des passions applicable à la pratique des arts plastiques.* Cited in Crawford (2021). *Atlas of AI*, p. 158.

39 Crawford (2021). *Atlas of AI*, pp. 158, 162–164.

40 Becky Little. 2019. What type of criminal are you? 19th century doctors claimed to know by your face. *History*, 8 August.

41 Ibid. This refers to Xiaolin Wu and Xi Zhang. 2016. Automated inference on criminality using face images. Preprint, arXiv (13 November).

42 Stan Z. Li and Anil K. Jain (eds.). 2011. *Handbook of Face Recognition*, 2nd edn. Springer. Kelly A. Gates. 2011. *Our Biometric Future: Facial Recognition Technology and the Culture of Surveillance.* New York University Press.

43 Chinoy (2019). The racist history behind facial recognition.

44 Angela Saini. 2019. *Superior: The Return of Race Science*, pp. 68–71, 74–75, 77, 85, 88. London: Fourth Estate.

45 Chinoy (2019). The racist history behind facial recognition. See also 'Sir Francis Galton, eugenics and final years'. The Galton Institute.

46 Kashmir Hill. 2021. Your face is not your own. *New York Times Magazine*, 18 March. Kashmir Hill. 2021. What we learned about Clearview AI and its secret 'co-founder'. *New York Times*, 18 March.

47 Jana Winter. 2021. Facial recognition, fake identities and digital surveillance tools: inside the post office's covert internet operations program. *Yahoo! News*, 18 May.

48 Kashmir Hill. 2021. Clearview AI's facial recognition app called illegal in Canada. *New York Times*, 3 February.

49 Stephanie Bodoni. 2021. Clearview AI hit by wave of European privacy complaints. *Bloomberg*, 27 May.

50 Byron Kaye. 2021. Australia says US facial recognition software firm Clearview breached privacy law. *Reuters*, 3 November.

51 Ibid.

52 Ashlee Vance. 2016. A look inside Russia's creepy, innovative internet. *Hello World, Bloomberg,* November.

53 Drew Harwell. 2021. This facial recognition website can turn anyone into a cop – or a stalker. *Washington Post,* 14 May.

54 Alice Hines. 2021. How normal people deployed facial recognition on Capitol Hill protestors. *Vice,* 2 February. Jane Wakefield. 2020. PimEyes facial recognition website 'could be used by stalkers'. *BBC News,* 11 June.

55 Sidney Fussell. 2021. Baltimore may soon ban facial recognition for everyone but cops. *WIRED,* 18 June.

56 Os Keyes, Nikki Stevens and Jacqueline Wernimont. 2019. The government Is using the most vulnerable people to test facial recognition software. *Slate,* 17 March. Nikki Stevens and Os Keyes. 2021. Seeing infrastructure: race, facial recognition and the politics of data. *Cultural Studies* **35**(4–5), 833–853.

57 Crawford (2021). *Atlas of AI,* pp. 90–94, 109–110.

58 Ibid., pp. 89–91. As Peter Fussey observes, the 9/11 attackers' identities were not in question, it was their intention, and this would not have been detectable with facial recognition technology. The hijackers had checked in with their real names, after all. See Peter Fussey. 2007. Observing potentiality in the global city: surveillance and counterterrorism in London. In *International Criminal Justice Review* **17**(3), 171–192.

59 Brad Smith. 2018. Face recognition: it's time for action. Company blog, Microsoft, 6 December.

60 O'Neil (2015). *Weapons of Math Destruction.*

61 Bruce Schneier. 2020. We're banning facial recognition. We're missing the point. *New York Times,* 20 January.

62 Stephanie Kirchgaessner, Paul Lewis, David Pegg, Sam Cutler, Nina Lakhani and Michael Safi. 2021. Revealed: leak uncovers global abuse of cyber-surveillance weapon. *The Guardian,* 18 July.

63 Idemia has evolved thusly: a company called Safran became Safran Morpho, which became OT-Morpho, and it is now known as Idemia. See Accenture. 2015. Accenture unique identity services, p. 5.

64 Arpan Chaturvedi. 2018. The key arguments in Supreme Court against Aadhaar. *Bloomberg,* 21 March. Accenture. 2010. Unique Identification Authority of India

(UIDAI) selects Accenture to implement a multimodal biometric solution for 'Aadhaar' program. Press release, 28 July.

65 Nayantara Ranganathan. 2021. The economy (and regulatory practice) that biometrics inspires: a study of the Aadhaar project. In *Regulating Biometrics: Global Approaches and Urgent Questions*, AI Now, pp. 52–61.

66 Rachna Khaira. 2018. Rs 500, 10 minutes, and you have access to billion Aadhaar details. *The Tribune*, 3 January. Vindi Doshi. 2018. A security breach in India has left a billion people at risk of identity theft. *Washington Post*, 4 January.

67 Upmanyu Trivedi. 2018. World's largest ID database exposed by Indian Government errors. *Bloomberg*, 14 May.

68 Aman Sethi. 2018. Aadhaar seeding fiasco: how to geo-locate by caste and religion in Andhra Pradesh with one click. Is this Big Brother enough for you? *Huffington Post*, 25 April.

69 Madhumita Murgia. 2021. India deploys facial recognition surveilling millions of commuters. *Financial Times*, 26 August.

70 Andrew Kersley. 2021. Couriers say Uber's 'racist' facial identification technology got them fired. *WIRED*, 1 March.

71 Joy Buolamwini and Timnit Gebru. 2018. Gender shades: intersectional accuracy disparities in commercial gender classification. *Proceedings of Machine Learning Research* **81**, 1–15. Conference on Fairness, Accountability and Transparency.

72 Kersley (2021). Couriers say Uber's 'racist' facial identification technology got them fired.

73 Avi Asher-Schapiro. 2021. For this Amazon van driver, AI surveillance was the final straw. *Reuters*, 19 March.

74 Ibid.

75 Danielle Abril and Drew Harwell. 2021. Keystroke tracking, screenshots, and facial recognition: the boss may be watching long after the pandemic ends. *Washington Post*, 24 September.

76 Richard Baimbridge. 2021. Why your face could be set to replace your bank card. *BBC News*, 24 January.

77 Sam Dean. 2020. Forget credit cards – now you can pay with your face. Creepy or cool? *Los Angeles Times*, 14 August.

78 Yuan Yang and Madhumita Murgia. 2019. Facial recognition: how China cornered the surveillance market. *Financial Times*, 6 December.

79 Cynthia O'Murchyu. 2021. Facial recognition cameras arrive in school canteens. *Financial Times*, 16 October.

80 Along with Pippa King from Biometrics in Schools and Jen Persson of Defend Digital Me, the author spoke with Lord Scriven about this ahead of the House of Lords debate.

81 House of Lords. 2021. Biometric recognition technologies in schools. *Hansard*, volume 815 (debated on Thursday 4 November).

82 Ginia Bellafante. 2019. The landlord wants facial recognition in its rent-stabilized buildings. Why? *New York Times*, 28 March.

83 Yasmin Gagne. 2019. How we fought our landlord's secretive plan for facial recognition – and won. *Fast Group*, 22 November. Mutale Nkonde. 2019–2020. Automated anti-blackness: facial recognition in Brooklyn, New York. In *Harvard Kennedy School Journal of African American Policy 'Anti-Blackness in Policy Making: Learning from the Past to Create a Better Future'*, pp. 30–36.

84 Ibid.

85 Claudia Garcia-Rojas interview with Simone Browne. 2016. The surveillance of blackness: from the trans-Atlantic Slave Trade to contemporary surveillance technologies. *Truthout*, 3 March. See also Simone Browne. 2015. *Dark Matters: On the Surveillance of Blackness*. Durham, NC/London: Duke University Press.

86 Elise Thomas. 2018. Logged, tracked and in danger: how the Rohingya got caught in the UN's risky biometric database. *WIRED*, 3 December.

87 Ben Hayes and Massimo Marelli. 2020. Reflecting on the international committee of the Red Cross's biometric policy minimizing centralized databases. In *Regulating Biometrics: Global Approaches and Urgent Questions*, edited by Amba Kak, pp. 70–77. AI Now Institute.

88 James Eaton-Lee. 2021. Oxfam's new policy on biometrics explores safe and responsible data practice. Oxfam, 24 June.

89 Jo Burton. 2021. 'Doing no harm' in the digital age: what the digitalization of cash means for humanitarian action. *IRRC* **913**, March. Citing Ben Hayes and Massimo Marelli. 2019. Facilitating innovation, ensuring protection: the ICRC biometrics policy. Humanitarian Law and Policy Blog, 18 October.

90 Hayes and Marelli (2019). Reflecting on the International Committee of the Red Cross's Biometric Policy Minimizing Centralized Databases, p. 70, footnote 2.

91 Ibid., p. 77.

92 Robert Channick. 2021. Waiting for your $345 from the Illinois Facebook privacy settlement? Here's why it's delayed. *Chicago Tribune*, 5 April.

93 Woodrow Hartzog. 2020. BIPA: the most important biometric privacy law in the US? In *Regulating Biometrics: Global Approaches and Urgent Questions*, edited by Amba Kak, p. 101, footnote 31. AI Now Institute.

94 David Ingram. 2021. Facebook to delete 1 billion people's 'facial recognition templates'. *NBC News*, 2 November.

95 Ibid., pp. 101, 96, footnote 1. Clearview AI scraped billions of images of people without their permission from social media websites to power their facial recognition app. Clearview filed legal documents in Illinois stating that 'Clearview is cancelling the accounts of every customer who was not either associated with law enforcement or some other federal, state, or local government department, office, or agency.'

96 *ACLU v. Clearview AI*. Lawsuit filed on 28 May 2020: see www.aclu.org/cases/aclu-v-clearview-ai.

97 Matt Burgess. 2020. Co-op is using facial recognition tech to scan and track shoppers. *WIRED*, 12 October.

98 Fionntán O'Donnell. 2018. A brief guide to BBC R&D's video face recognition. *BBC News*, 27 February.

99 Ilya Arkhipov and Jake Rudnitsky. 2021. In Moscow, Big Brother is watching and recognising protestors. *Bloomberg*, 2 May.

100 Ibid.

101 Ibid.

102 Paul Mozur. 2019. In Hong Kong protests, faces become weapons. *New York Times*, 26 July.

103 Clare Garvie. 2019. You're in a police line-up right now. *New York Times*, 15 October. Clare Garvie testimony before the US Congress, 22 May 2019.

104 Clare Garvie. 2021. Interview with Anderson Cooper. Police departments adopting facial recognition tech amid allegations of wrongful arrests. *CBS: 60 Minutes,* 16 May.

105 Steven Feldstein. 2019. The global expansion of AI surveillance. Carnegie Endowment for International Peace, 17 September. Cited in Yang and Murgia (2019). Facial recognition: how China cornered the surveillance market. See also Paul Mozur, Jonah M. Kessel and Melissa Chan. 2019. Made in China, exported to the world: the surveillance state. *New York Times,* 24 April.

106 Ross Anderson. 2020. The panopticon is already here. *The Atlantic,* September.

107 Anna Gross, Madhumita Murgia and Yuan Yang. 2019. Chinese tech groups shaping UN facial recognition standards. *Financial Times,* 1 December.

108 Ryan Mac. 2021. Facebook apologizes after AI puts 'primates' label on video of black men. *New York Times,* 3 September.

109 'The Met currently uses NEC's NeoFace Live Facial Recognition technology to take images and compare them to images of people on the watchlist. It measures the structure of each face, including distance between eyes, nose, mouth and jaw to create a facial template.' See London Metropolitan Police: 'Live facial recognition'.

110 Ibid.

111 UK Department for International Trade. 2021. Export Control Joint Unit. Exporting military or dual-use technology: definitions and scope. 22 March. Rashida Richardson. 2021. Facial recognition in the public sector: the policy landscape. German Marshall Fund, 3 February. Els J. Kindt. 2021. Transparency and accountability mechanisms for facial recognition. German Marshall Fund, 3 February.

112 Noah Shachtman. 2010. Army reveals Afghan biometric plan; millions scanned, carded by May. *WIRED,* 24 September. Greg Jaffe. 2011. An unusual mission in Afghanistan and the troops who suffered to carry it out. *Washington Post,* 13 January.

113 Thom Shanker. 2011. To track militants, US has a system that never forgets a face. *New York Times,* 13 July. Noah Shachtman. 2007. Iraq's biometric database could become 'hit list': Army. *WIRED,* 15 August.

114 Rina Chandran. 2021. Afghans scramble to delete digital history, evade biometrics. *Reuters,* 18 August.

115 Thom Patterson. 2018. US airport opens first fully biometrics terminal. *CNN,* 3 December. Citing US Customs and Border Protection, 'Biometrics' and 'Biometrics FAQ' on privacy and data retention policies.

116 Lori Arantani. 2018. Facial-recognition scanners at airports raise privacy concerns. *Washington Post,* 15 September.

117 Emily Birnbaum. 2019. DHS wants to use facial recognition on 97 percent of departing air passengers by 2023. *The Hill,* 18 April. Davey Alba. 2019. The US government will be scanning your face at 20 top airports, documents show. *BuzzFeed News,* 11 March.

118 The White House. 2021. Fact sheet: President Biden sends Immigration Bill to Congress as part of his commitment to modernize our immigration system. 20 January. Niamh Kinchin. 2021. AI facial analysis is scientifically questionable. Should we be using it for border control? *The Conversation,* 23 February.

119 Accenture Border Services. Accenture. 2015. Accenture Unique Identity Services. Accenture only did the implementation; the technology was from a French/US conglomerate called Idemia for the UK, US and EU cases.
120 Partnership between Accenture, UK Border Agency, BAA and various suppliers.
121 Ibid.
122 In 2013 the UNHCR commissioned Accenture to build this system. It captures fingerprints, iris and facial data.
123 Accenture. 2015. Video analytics: operational, marketing and security insights from CCTV. 23 May.
124 In 2018 Accenture, along with Capgemini and Deloitte, won this contract. Kat Hall. 2018. Accenture, Capgemini, Deloitte creating app to register 3m EU nationals living in Brexit Britain. *The Register*, 9 April.
125 Maryam Ahmed. 2020. UK passport photo checker shows bias against dark-skinned women. *BBC News*, 8 October.
126 John Dunne. 2019. Man stunned as passport photo check sees lips as open mouth. *Evening Standard*, 19 September.
127 BBC. 2019. Passport facial recognition checks fail to work with dark skin. *BBC News*, 9 October.
128 David Green. 2021. NuraLogix's work on public surveillance tools raises concerns about cooperating with firms tied to Chinese government. *Globe and Mail*, 16 April.
129 Crawford (2021). *Atlas of AI*, pp. 123–150. Saini (2019). *Superior*.
130 *The Economist*. 2017. Nowhere to hide: what machines can tell from your face. *The Economist*, 9 September.
131 Kelion (2021). Huawei patent mentions use of Uighur-spotting tech. Avi Asher-Schapiro. 2021. China found using surveillance firms to help write ethnic tracking specks. *Reuters*, 30 March.
132 George Joseph and Kenneth Lipp. 2018. IBM used NYPD surveillance footage to develop technology that lets police search by skin colour. *The Intercept*, 6 September. George Joseph. 2019. Inside the video surveillance program IBM built for Philippine strongman Rodrigo Duterte. *The Intercept*, 20 March.
133 Daniel Berehulak. 2016. They are slaughtering us like animals. *New York Times*, 7 December. See also Joseph (2019). Inside the video surveillance program IBM built for Philippine strongman Rodrigo Duterte.
134 Bobby Allyn. 2020. IBM abandons facial recognition products, condemns racially biased surveillance. *NPR*, 9 June.
135 Crawford (2021). *Atlas of AI*, p. 115.
136 *The Economist*. 2017. Advances in AI are used to spot signs of sexuality. *The Economist*, 9 September. Derek Hawkins. 2017. Researchers use facial recognition to predict sexual orientation. LBGT groups aren't happy. *Washington Post*, 12 September. Blaise Aguera y Arcas, Alexander Todorov and Margaret Mitchell. 2018. Do algorithms reveal sexual orientation or just expose our stereotypes? *Medium*, 11 January. *The Guardian*. 2018. 'I was shocked it was so easy': meet the professor who says facial recognition can tell if you're gay. *The Guardian*, 7 July. Adrianne Jeffries. 2017. That study on artificially intelligent 'gaydar' is now under ethical review. *The Outline*, 11 September.
137 BBC Reality Check team. 2021. Homosexuality: the countries where it is illegal to be gay. *BBC News*, 12 May.

138 Khari Johnson. 2021. DeepMind researchers say AI poses a threat to people who identify as queer. *VentureBeat,* 18 February.

139 Michal Kosinski. 2021. Facial recognition technology can expose political orientation from naturalistic facial images. *Nature,* 11 January. John Naughton. 2021. Can facial recognition really reveal political orientation? *The Guardian,* 23 January.

140 Madhumita Murgia. 2021. Emotion detection: can AI detect human feelings from a face? *Financial Times,* 12 May. Leo Kelion. 2019. Emotion-detecting tech should be restricted by law – AI Now. *BBC News,* 12 December.

141 Crawford (2021). *Atlas of AI,* pp. 151–152.

142 Drew Harwell. 2019. A face-scanning algorithm increasingly decides whether you deserve the job. *Washington Post,* 6 November.

143 Will Knight. 2021. Job screening service halts facial analysis of applicants. *WIRED,* 12 January. Roy Maurer. 2021. HireVue discontinues facial analysis screening. *SHRM,* 3 February.

144 Joy Buolamwini. 2019. Artificial intelligence has a problem with gender and racial bias. *TIME,* 7 February. Buolamwini and Gebru (2018). Gender shades: intersectional accuracy disparities in commercial gender classification.

145 National Institute of Standards and Technology. 2019. NIST study evaluates effects of race, age, sex on face recognition software. 19 December. Patrick Grother, Mei Ngan and Kayee Hanaoka. 2018. Ongoing face recognition vendor test (FRVT) part 2: identification. NIST/US Department of Commerce, November. Patrick Grother, Mei Ngan and Kayee Hanaoka. 2019. Face recognition vendor test (FRVT) part 3: demographic effects. NIST/US Department of Commerce, December. Drew Harwell. 2019. Federal study confirms racial bias of many facial-recognition systems, casts doubt on their expanding use. *Washington Post,* 19 December.

146 Jamiles Lartey. 2015. By the numbers: US police kill more in days than other countries do in years. *The Guardian,* 9 June.

147 Drew Harwell. 2021. Wrongfully arrested man sues Detroit police over false facial recognition. *Washington Post,* 13 April.

148 Andrew Westrope. 2019. Axon decides against facial recognition for body cameras. *GovTech Biz,* 27 June. Axon. 2019. First report of the Axon AI and Policing Technology Ethics Board. June. Brian Brackeen. 2018. Facial recognition software is not ready for use by law enforcement. *TechCrunch,* 25 June.

149 Olivia Solon. 2019. Facial recognition's 'dirty little secret': millions of online photos scrapped without consent. *NBC,* 12 March.

150 Concerned Researchers (a list). 2019. On recent research auditing commercial facial analysis technology. *MIT Media Lab,* 3 April. Dina Bass. 2019. Amazon schooled on AI facial technology by Turing Award winner. *Bloomberg,* 3 April. Natasha Singer and Cade Metz. 2019. Many facial recognition systems are biased, says US study. *New York Times,* 19 December.

151 Kara Swisher interview with Joy Buolamwini. 2021. She's taking Jeff Bezos to task. *New York Times,* 19 April.

152 Javier Espinoza and Madhumita Murgia. 2020. Sundar Pichai supports calls for moratorium on facial recognition. *Financial Times,* 20 January.

153 Stephanie Hare. 2018. We must face up to the threat posed by biometrics. *Financial Times,* 8 August.

154 Drew Harwell. 2021. Senators seek limits on some facial recognition use by police, energizing surveillance technology debate. *Washington Post,* 21 April.

155 Drew Harwell. 2021. Amazon extends ban on police use of its facial recognition technology indefinitely. *Washington Post,* 18 May.

156 Clare Garvie, Alvaro Bedoya and Jonathan Frankle. 2016. Unregulated police face recognition in America. *Perpetual Line-Up,* 18 October. Clare Garvie. 2019. Garbage in, garbage out: face recognition on flawed data. Georgetown Law, 16 May. Clare Garvie and Laura M. Moy. 2019. America under watch: face surveillance in the United States. Georgetown Law, 16 May. Clare Garvie interviewed by Anderson Cooper. 2021. Police departments adopting facial recognition tech amid allegations of wrongful arrests. *CBS: 60 Minutes,* 16 May.

157 Lauren Bridges. 2021. Amazon's Ring is the largest civilian surveillance network the US has ever seen. *The Guardian,* 18 May. Leo Kelion. 2020. Why Amazon knows so much about you. *BBC News.* BBC. 2020. Amazon: what they know about us (documentary). *BBC Panorama,* 22 February.

158 Stephanie Hare. 2018. Facial recognition. *Tech Tent, BBC World Service,* 21 September.

159 Jon Sharman. 2018. Metropolitan Police's facial recognition technology 98% inaccurate, figures show. *Independent,* 13 May. Annabelle Dickson. 2019. UK looks the other way on AI. *Politico,* 11 March.

160 Big Brother Watch. 2018. Face off: the lawless growth of facial recognition in UK policing. May.

161 Peter Fussey and Daragh Murray. 2020. Policing uses of live facial recognition in the United Kingdom. In *Regulating Biometrics: Global Approaches and Urgent Questions,* edited by Amba Kak. AI Now Institute.

162 Peter Fussey and Daragh Murray. 2019. New report ['Independent report of the London Metropolitan Police Service's trial of live facial recognition technology'] raises concerns over Met Police trials of live facial recognition technology. University of Essex, 3 July (www.essex.ac.uk/news/2019/07/03/met-police-live-facial-recognition-trial-concerns). National Physical Laboratory. 2020. Metropolitan Police Service live facial recognition trials. Report, February.

163 Fussey and Murray (2019). Independent report of the London Metropolitan Police Service's trial of live facial recognition technology.

164 Haroon Siddique. 2016. Who was Jean Charles de Menezes? *The Guardian,* 30 March.

165 Anonymous. 2020. As a black police officer, I know the Met is still institutionally racist. *The Guardian,* 15 June. Calum Leslie. 2020. Police racism inquiries in the UK: do they change how things work? *BBC News,* 10 August. Adam Vaughn. 2021. UK still using racially biased passport tool despite available update. *New Scientist,* 15 March. Adam Vaughn. 2019. UK launched passport photo checker it knew would fail with dark skin. *New Scientist,* 9 October. Amelia Gentleman. 2019. *The Windrush Betrayal: Exposing the Hostile Environment.* London: Guardian Faber Publishing. UK National Audit Office. 2021. Home Office investigation into the Windrush compensation scheme. 21 May. Jermaine Jenas. 2021. 'I was 10 when it first happened': Jermaine Jenas on why stop-and-search is failing black children. *The Guardian,* 28 May. Vikram Dodd and Dan Sabbagh.

2021. Daniel Morgan murder: inquiry brands Met police 'institutionally corrupt'. *The Guardian*, 15 June.

166 Sarah Turnnidge. 2021. Police still can't explain 'unfair' use of powers against BAME people, watchdog says. *Huffington Post*, 26 February. Sarah Turnnidge. 2021. Exclusive: the Met Police are more likely to publish your mugshot if you're black. *Huffington Post*, 11 March.

167 BBC Click. 2019. Fines and facial recognition. *BBC News*, 17 May.

168 The London Metropolitan Police's statement about the issuing of this fine is here: 'Trial of facial recognition software, Freedom of information request reference no: 01.FOI.19.011711'.

169 Fussey and Murray (2020). Policing uses of live facial recognition in the United Kingdom. See also Court of Appeal (CA) judgement in the case of R (on the application of Edward BRIDGES) v The Chief Constable of South Wales Police, 11 August 2020.

170 Ibid.

171 UK Information Commissioner's Office. 2021. ICO urges caution in using facial recognition in public places. Blog post, 18 June.

172 UK Biometrics Commissioner Professor Paul Wiles. 2018. Facial recognition. *Tech Tent*, *BBC World Service*, 21 September.

173 Alexi Mostrous. 2019. Ministers must act of 'explosion' of spy technology. *The Times*, 17 June.

174 Paul Wiles: 'At present there are 23 million images in the Police National Database and 10 million searchable facial images, but this does not necessarily mean 10 million people because some of the images will be duplicates.' From Sebastian Klovig Skelton. 2019. UK police should not deploy live facial recognition technology until issues are resolved, MPs told. *Computer Weekly*, 22 March. Written evidence submitted by Steve Wood, Deputy Commissioner for Policy, Information Commissioner's Office (WBC0008), 19 March 2019.

175 Nick Hopkins and Jake Morris. 2015. 'Innocent people' on police photos database. *BBC News*, 3 February.

176 Stephanie Hare. 2018. Facial recognition. *Tech Tent*, *BBC World Service*, 21 September.

177 House of Commons. 2019. Facial recognition and the biometrics strategy. Volume 659: debated on Wednesday 1 May.

178 Commons Select Committee. 2019. Issues with biometrics and forensics significant risk to effective functioning of the criminal justice system. 18 July.

179 BBC. 2021.New police CCTV use rules criticised as bare bones. *BBC News*, 17 August.

180 'The Met has used the technology several times since 2016, according to its website, including at Notting Hill Carnival in 2016 and 2017, Remembrance Day in 2017 and Port of Hull docks assisting Humberside Police last year.' From Sam Tobin. 2019. What is facial recognition technology and why is it controversial? *Belfast Telegraph*, 4 September. Jason Douglas and Parmy Olson. 2020. London police to start using facial-recognition cameras. *Wall Street Journal*, 24 January. Leicestershire police were one of three forces using it in the UK in 2018 along

with the Met and South Wales police: see Patricia Nilsson. 2018. How UK police are using facial recognition software. *Financial Times*, 12 October.

181 Madhumita Murgia. 2019. How one London wine bar helped Brazil to cut crime. *Financial Times*, 8 February.

182 Mark Townsend. 2019. Police forces halt trials of facial recognition systems. *The Guardian*, 17 August.

183 Joe Purshouse and Liz Campbell. 2019. Before facial recognition tech can be used, it needs to be limited. *Independent*, 21 February.

184 Jenny Rees. 2020. Facial recognition use by South Wales Police ruled unlawful. *BBC News*, 11 August.

185 South Wales Police. 2021. New facial recognition mobile app to identify vulnerable, missing and wanted individuals. Press release, 7 December.

186 BBC. 2020. Facial recognition: 'No justification' for Police Scotland to use technology. *BBC News*, 11 February.

187 Freedom of Information Request. Request number F-2018-01186. Subject: Facial Recognition. Freedom of Information Request. Request number F-2018-02564. Subject: Facial recognition.

188 Freedom of Information Request. Request number F-2020-00188. Subject: Facial Recognition Technology.

189 Hao (2021). This is how we lost control of our faces.

190 Madhumita Murgia. 2021. Police should not be banned from using facial recognition technology, says watchdog. *Financial Times*, 3 May.

191 'Automated Facial Recognition Technology (Moratorium and Review) Bill [HL]', Private Members Bill (starting in the House of Lords), originated in the House of Lords, session 2019-19.

192 Surveillance Commissioner. 2018. Surveillance Camera Commissioner Annual Report 2016/17, pp. 12–13. January. James Temperton. 2015. One nation under CCTV: the future of automated surveillance. *WIRED*, 17 August. On 9 March 2021 the Home Secretary Priti Patel appointed Fraser Sampson as the government's new independent Biometrics and Surveillance Commissioner, merging into one what had been two separate roles held by Professor Paul Wiles and Tony Porter, respectively. See Home Office (UK Government) website. 2021. New Biometrics and Surveillance Camera Commissioner appointed. 9 March.

193 Madhumita Murgia. 2019. How London became a test case for using facial recognition in democracies. *Financial Times*, 1 August. Surveillance Commissioner. 2018. Surveillance Camera Commissioner Annual Report 2016/17, pp. 12–13. January.

194 James Vincent. 2020. London police to deploy facial recognition cameras across the city. *The Verge*, 24 January.

195 Ibid. Citing Tony Porter (then UK Surveillance Camera Commissioner). 2019. Surveillance Camera Commissioner: annual report 2017–2018, 22 January. See also Murgia (2019). How one London wine bar helped Brazil to cut crime.

196 Big Brother Watch. 2019. Facial recognition 'epidemic' in the UK. 16 August. BBC. 2015. Facewatch 'thief recognition' CCTV on trial in UK stores. *BBC News*, 16 December.

197 Mark Blunden. 2019. Face-scanning CCTV approved for London hot-spots despite privacy storm. *Evening Standard*, 2 September.

198 Kari Paul. 2019. Amazon's doorbell camera Ring is working with police – and controlling what they say. *The Guardian*, 30 August. London Metropolitan Police. 2020. Amazon Ring internet-connected camera-enabled doorbells. Freedom of information request reference no: 01.FOI.19.012185, January. Privacy International. 2020. One ring to watch them all. 25 June. Sam Wollaston. 2021. I spy: are smart doorbells creating a global surveillance network? *The Guardian*, 26 June.

199 Megan Wollerton. 2021. The best facial recognition security cameras of 2021. *CNET*, 5 May.

200 Associated Press. 2019. Amazon has considered facial recognition in its Ring doorbells. *Market Watch*, 19 November.

201 Madhumita Murgia. 2019. London's King's Cross uses facial recognition in security cameras. *Financial Times*, 12 April. Zoe Kleinman. 2019. King's Cross face recognition 'last used in 2018'. *BBC News*, 2 September.

202 Stephanie Hare. 2019. Biometrics: facial recognition on trial. CogX Festival of AI, 11 June.

203 Madhumita Murgia. 2019. New details emerge of King's Cross facial recognition plans. *Financial Times*, 3 September. Leo Kelion. 2019. Met Police gave images for King's Cross facial recognition scans. *BBC News*, 6 September.

204 Kevin Rawlinson. 2019. ICO opens investigation into use of facial recognition in King's Cross. *The Guardian*, 15 August.

205 Helen Warrell and Nic Fildes. 2021. UK spy agencies push for curbs on Chinese 'smart city' technology. *Financial Times*, 18 March. Gordon Corera. 2021. Spy bosses warn of cyber-attacks on smart cities. *BBC News*, 7 May. Gordon Corera. 2021. GCHQ chief warns of tech 'moment of reckoning'. *BBC News*, 23 April.

206 David Shepardson. 2021. Five Chinese companies pose threat to US national security – FCC. *Reuters*, 13 March. Demetri Sevastopulo. 2021. Washington to bar US investors from 59 Chinese companies. *Financial Times*, 4 June.

207 James Kynge and Demetri Sevastopulo. 2019. US pressure building on investors in China surveillance group. *Financial Times*, 29 March. Camilla Hodgson. 2019. 2,000 Chinese-made surveillance cameras across US government. *Financial Times*, 30 July.

208 Chris Buckley, Paul Mozur and Austin Ramzy. 2019. How China turned a city into a prison. *New York Times*, 4 April. John Sudworth. 2018. China's hidden camps: what's happened to the vanished Uighurs of Xinjiang? *BBC Online*, 24 October. Chris Buckley and Austin Ramzy. 2018. China's detention camps for Muslims turn to forced labor. *New York Times*, 16 December.

209 Ryan Gallagher. 2019. Hikvision cameras linked to Chinese Government stir alarm in UK Parliament. *The Intercept*, 8 April.

210 Avi Asher-Schapiro. 2021. Exclusive – half London councils found using Chinese surveillance tech linked to Uighur abuses. *Reuters*, 18 February. Gallagher (2019). Cameras linked to Chinese Government stir alarm in UK Parliament.

211 Ibid.

212 Ibid.

213 Mark Gilbert. 2021. Your face is the next frontier in ESG investing. *Bloomberg Opinion*, 21 June.

NOTES

214 Campbell *et al.* (2020). Sci-fi surveillance: Europe's secretive push into biometric technology.
215 Bentham's brother Samuel had the idea but it was Jeremy who took it forward and got it built. See Thomas McMullan. 2015. What does the panopticon mean in the age of digital surveillance? *The Guardian*, 23 July.
216 Jeremy Bentham. 1843. *The Works of Jeremy Bentham*, volume 4, p. 40. Edinburgh: William Tate. Nick Harkaway. 2017. Who's looking at you? *BBC Radio 4*, 7 October.
217 Michel Foucault. 1975. *Surveiller et Punir*. Paris: Gallimard. Michel Foucault. 1991. *Discipline and Punish*, translated by Alan Sheridan (see chapter on 'Panopticism'). London: Penguin.
218 European Commission. 2021. Proposal for a regulation laying down harmonised rules on artificial intelligence (Artificial Intelligence Act). 21 April.
219 Ibid. See also Javiar Espinoza and Madhumita Murgia. 2021. EU lawmakers propose strict curbs on use of facial recognition. *Financial Times*, 20 April.
220 Melissa Heikkilä. 2021. Europe's AI rules open door to mass use of facial recognition, critics warn. *Politico*, 7 June.
221 Sebastian Klovig Skelton. 2021. Europe's proposed AI regulation falls short on protecting rights. *Computer Weekly*, 14 June. Kevin Cahill. 2018. New controversies as Trump passes US Foreign Intelligence Surveillance Act. *Computer Weekly*, 23 January.
222 Pete Fussey, Bethan Davies and Martin Innes. 2021. 'Assisted' facial recognition and the reinvention of suspicion and discretion in digital policing. *British Journal of Criminology* 61(2), 325–344.
223 Catherine Thorbecke. 2021. Citing human rights risks, UN calls for ban on certain AI tech until safeguards are set up. *ABC News*, 15 September.
224 Ed Markey. 2021. Senators Markey, Merkley lead colleagues on legislation to ban government use of facial recognition, other biometric technology. Press release, 15 June. So far efforts to regulate or ban facial recognition technology at the federal level have failed. See: Senators Cory Booker (D-New Jersey) and Jeff Merkley (D-Oregon) proposed federal facial recognition moratorium. 2020. The Ethical Use of Facial Recognition Act. February. Senators Brian Schatz (D-Hawaii) and Roy Blunt (R-Missouri) legislation in 2019 to require companies to gain people's consent before using facial recognition on them in public places and sharing their data with third parties. Senators Christopher Coons (D-Del) and Mike Lee (R-Utah)'s bill to require law enforcement to obtain court orders to use facial recognition software for extended surveillance. All of these were reported in Chris Mills Rodrigo. 2020. Booker, Merkley propose federal facial recognition moratorium. *The Hill*, 12 February.
225 'Automated Facial Recognition Technology (Moratorium and Review) Bill [HL]', Private Members' Bill (starting in the House of Lords). Originated in the House of Lords, session 2019-19.
226 Peter Walker, Heather Stewart and Haroon Siddique. 2021. More than 2m voters may lack photo ID required under new UK bill. *The Guardian*, 11 May. Robin Wright and Sebastian Payne. 2021. UK's voter ID plan 'an expensive distraction'. *Financial Times*, 13 May.

227 Stephanie Hare. 2021. Could Covid-19 vaccine passports use biometrics? *BBC Click*, 20 April. Cristina Criddle. 2021. NHS app ready to become vaccine passport next week. *BBC News*, 11 May.

Chapter 4: Pandemic? There's an app for that

1   World Health Organization: WHO COVID-19 dashboard.
2   As per the British system of devolved government, England, Wales, Scotland and Northern Ireland have their own health systems, which meant that lockdowns and various other pandemic responses varied throughout the United Kingdom.
3   House of Commons Library. 2021. A history of English lockdown laws. 30 April.
4   BBC. 2020. Coronavirus: police get access to NHS Test and Trace self-isolation data. *BBC News*, 18 October.
5   Jennifer Valentino-DeVries. 2020. Coronavirus apps show promise but prove a tough sell. *New York Times*, 7 December.
6   Apple/Google. 2020. Contact tracing: Bluetooth specification, preliminary – subject to modification and extension. April. Rory Cellan-Jones. 2021. *Always On: Hope and Fear in the Smartphone Era.* London: Bloomsbury.
7   John Drake. 2021. How 'digital contact tracing' reduced Covid-19 cases in the UK by 20%. *Forbes*, 21 May.
8   Helene Fouquet. 2020. France says Apple bluetooth policy is blocking virus tracker. *Bloomberg*, 20 April.
9   Joe Miller and Leila Abboud. 2020. German U-turn over coronavirus tracking app sparks backlash. *Financial Times*, 27 April. Siddarth Venkataramakrishnan. 2020. Contact-tracing apps must not be used for mass surveillance, warn experts. *Financial Times*, 20 April.
10 Laura Spinney. 2020. Germany's COVID-19 expert: 'For many, I'm the evil guy crippling the economy'. *The Guardian,* 26 April. Cellan-Jones (2021). *Always On*, pp. 232–233.
11 Susan Landau. 2021. *People Count: Contact-Tracing Apps and Public Health*, p. 61. Cambridge, MA: MIT Press.
12 Matthew Weaver. 2018. Minister forced to change his own app after data-mining complaints. *The Guardian,* 22 March. Matt Burgess. 2018. Matt Hancock MP has launched an app. And he wants all your data. *WIRED,* 1 February.
13 Madhumita Murgia. 2019. Home Office app for EU citizens easy to hack. *Financial Times,* 14 November. Promon. 2019. The Home Office's Brexit app lacks basic security, allowing hackers to steal passport information and facial IDs. Press Release, 14 November. In 2018 Accenture, along with Capgemini and Deloitte, won this contract: see Kat Hall. 2018. Accenture, Capgemini, Deloitte creating app to register 3m EU nationals living in Brexit Britain. *The Register,* 9 April.
14 BBC. 2016. TalkTalk fined £400,000 for theft of customer details. *BBC News*, 5 October. Joanna Taylor. 2020. People want to know why a disgraced TalkTalk CEO is leading the government's testing programme. *Indy100,* 27 May.

15 Rory Cellan-Jones. 2020. Coronavirus contact tracing: my new skill. *BBC News*, 22 May.

16 UK Parliament Public Accounts Committee. 2021. 'Unimaginable' cost of Test & Trace failed to deliver central promise of averting another lockdown. 10 March. Nick Triggle. 2021. Covid-19: Test and Trace weaknesses remain, says watchdog. *BBC News*, 25 June.

17 Alex Hern. 2020. How Excel may have caused loss of 16,000 Covid tests in England. *The Guardian*, 5 October. BBC. 2020. Covid: test error 'should never have happened' – Hancock. *BBC News*, 5 October.

18 Thiemo Fetzer and Thomas Graeber. 2020. Does contact tracing work? Quasi-experimental evidence from an Excel error in England. *MedRxiv*, 15 December.

19 Rachel Schraer. 2020. Coronavirus: government apologises over tests shortages. *BBC News*, 8 September.

20 Devi Sridhar. 2020. Continual lockdowns are not the answer to bringing Covid under control. *The Guardian*, 10 October.

21 Sarah Boseley. 2020. Less than 20% of people in England self-isolate fully, Sage says. *The Guardian*, 11 September.

22 Nick Triggle. 2020. Coronavirus: self-isolation payment for low-income workers. *BBC News*, 27 August. BBC Panorama. 2020. Test and trace: 'I spoke to one person in four months'. *BBC News*, 28 September.

23 Article. 2020. Covid-19: up to £10,000 fine for failure to self-isolate in England. *BBC News*, 28 September.

24 Ibid.

25 Lucy Yardley. 2020. Interview with Andrew Marr. *Andrew Marr Show*, BBC, 18 October

26 Leo Kelion and Rory Cellan-Jones. 2020. NHS Covid-19 app: how England and Wales's contact tracing service works. *BBC News*, 23 September. Leo Kelion and Rory Cellan-Jones. 2020. NHS Covid-19 app: one million downloads of contact tracer for England and Wales. *BBC News*, 24 September.

27 Ibid.

28 Tom Calver and Gabriel Pogrund. 2020. Software bungle meant NHS Covid app failed to warn users to self-isolate. *Sunday Times*, 1 November.

29 UK Department of Health and Social Care. 2020. New payment for people self-isolating in highest risk areas. Press Release, 27 August. BBC. 2020. Coronavirus: self-isolation payment for low-income workers. *BBC News*, 27 August.

30 Prime Minister's Office 10 Downing Street. 2020. New package to support and enforce self-isolation. 20 September.

31 Josh Halliday. 2021. Majority in England turned down for self-isolation support, data shows. *The Guardian*, 2 February.

32 Rory Cellan-Jones. 2020. Covid-19 app users can't get isolation payment. *BBC News*, 23 October. Leo Kelion. 2020. Coronavirus: NHS Covid-19 app starts offering self-isolate payments. *BBC News*, 10 December.

33 Halliday (2021). Majority in England turned down for self-isolation support, data shows.

34 Josh Halliday. 2021. England's poorest areas hit by Covid 'perfect storm' – leaked report. *The Guardian*, 17 February.

35 BBC. 2021. Covid-19: few people with symptoms are self-isolating, study finds. *BBC News*, 1 April.

36 Ibid.

37 Ibid.

38 Robert Booth. 2021. English councils refuse 6 in 10 requests for self-isolation pay. *The Guardian*, 18 June.

39 Andrew Woodcock. 2021. Self-isolation payments held down to deter 'gaming' of system, Matt Hancock reveals. *Independent*, 10 June.

40 David Pegg. 2021. Fifth of UK Covid contracts 'raised red flags for possible corruption'. *The Guardian*, 22 April. BBC. 2021. Timeline: Covid contracts and accusations of 'chumocracy'. *BBC News*, 20 April.

41 Ella Wills. 2021. Covid: parts of England to trial self-isolation support. *BBC News*, 24 May.

42 UK Authority. 2020. Covid-19 app developers aim for overseas interoperability. Web page, UK Authority, 12 October.

43 BBC. 2020. UK contact-tracing apps start to talk to each other. *BBC News*, 5 November.

44 Naomi O'Leary. 2020. Irish, German and Italian contact-tracing apps to be linked in bid for 'safer' travel. *Irish Times*, 29 September.

45 Nicole Kobie. 2020. The race is on to make contact tracing apps work across borders. *WIRED*, 19 October.

46 Madhumita Murgia. 2021. Tim Spector: the data explorer who uncovered vital clues to Covid. *FT Magazine,* 31 July.

47 Selena Simmons-Duffin. 2021. How a gay community helped the CDC spot a COVID outbreak — and learn more about Delta. *NPR*, 6 August.

48 Ada Lovelace Institute. 2020. Exit through the App store? Rapid evidence review. Case study, April 20. Ada Lovelace Institute. 2020. International monitor: public health identity systems. Summary, June 22.

49 Dan Sabbagh and Alex Hern. 2020. UK abandons contact-tracing app for Apple and Google model. *The Guardian*, 18 June.

50 At the end of September 2020, France's app had had only 3 million downloads versus the UK's 12 million. Kim Willsher. 2020. French ministers in spotlight over poor take-up of 'centralised' Covid app. *The Guardian*, 29 September. Amanda Morrow. 2020. Reboot on the way for France's failed coronavirus tracking app. *RFI*, 13 October.

51 Tim Bradshaw. 2020. 2bn phones cannot use Google and Apple contact tracing tech. *Financial Times*, 20 April.

52 Rory Cellan-Jones and Zoe Kleinman. 2020. NHS Covid-19 app: 12 million downloads – and lots of questions. *BBC News*, 28 September.

53 BBC. 2020. Contact tracing apps can now work on older iPhones. *BBC News*, 15 December.

54 Leo Kelion. 2021. Covid-19: NHS app has told 1.7 million to self-isolate. *BBC News*, 9 February.

55 Tim Loh. 2020. Germany has its own Dr Fauci – and actually follows his advice. *Bloomberg Businessweek*, 28 September.

56 Zeynep Tufekci. 2020. This overlooked variable is the key to the pandemic. *The Atlantic.* 30 September.

57 Saira Asher. 2020. TraceTogether: Singapore turns to wearable contact-tracing
    Covid tech. *BBC News*, 4 July. John Geddie and Aradhana Aravindan. 2020.
    Singapore plans wearable virus-tracing device for all. *Reuters*, 5 June.
58 Ibid.
59 Ibid.
60 Reuters. 2021. Singapore COVID-19 contact-tracing data accessible to the police.
    Reuters, 4 January.
61 Robert Hinch, Will Probert, Anel Nurtay, Michelle Kendall, Chris Wymant,
    Matthew Hall, Katrina Lythgoe, Ana Bulas Cruz, Lele Zhao, Andrea Stewart,
    Luca Ferretti, Michael Parker, Daniel Montero, James Warren, Nicole K. Mather,
    Anthony Finkelstein, Lucie Abeler-Dörner, David Bonsall and Christophe Fraser.
    2020. Effective configurations of a digital contact tracing app: a report to NHS.
    Report, 16 April (updated 10 August 2020). Patrick Howell O'Neill. 2020. No
    coronavirus apps don't need 60% adoption to be effective. *MIT Technology
    Review*, 5 June.
62 Tim Bradshaw and Siddharth Venkataramakrishnan. 2021. NHS Covid app
    prevented 600,000 infections, claim researchers. *Financial Times*, 9 February.
63 Mark Briers, Chris Holmes and Christopher Fraser. 2021. Demonstrating
    the impact of the NHS COVID-19 app: statistical analysis from researchers
    supporting the development of the NHS COVID-19 app. Alan Turing
    Institute, 9 February. See also Chris Wymant *et al.* 2021. The epidemiological
    impact of the NHS COVID-19 app. *Nature* **594**, 408–412. Bradshaw and
    Venkataramakrishnan (2021). NHS Covid app prevented 600,000 infections, say
    researchers. BBC. 2021. NHS tracing app 'prevented thousands of deaths'. *BBC
    News*, 13 May.
64 Rory Cellan-Jones and Leo Kelion. 2021. Covid-19: NHS app has told 1.7 million to
    self-isolate. *BBC News*, 9 February.
65 Wymant *et al.* (2021). The epidemiological impact of the NHS COVID-19 App.
66 Ibid. See also Fabio Chiusi. 2021. Digital contact tracing apps: do they actually
    work? A review of early evidence. *AlgorithmWatch*, 8 July.
67 Tom Knowles. 2021. NHS Covid app usage at 'all-time high' despite claims users
    are deleting it. *The Times*, 7 July.
68 Chris Stokel-Walker and Lindsay Muscato. 2021. Is the UK's pingdemic good or
    bad? Yes. *MIT Tech Review*, 23 July. Michelle Kendall and Christophe Fraser. 2021.
    Evaluating epidemiological impacts of the NHS COVID-19 app: an August 2021
    update. Blog, 2 August.
69 GOV.UK. 2021. NHS COVID-19 app updated to notify fewer contacts to isolate.
    Press Release, 2 August. Jasmine Cameron-Chileshe. 2021. Number of alerts
    sent by NHS Covid app falls sharply. *Financial Times*, 5 August.
70 Craig Spencer. 2021. No, vaccinated people are not 'just as likely' to spread the
    coronavirus as unvaccinated people. *The Atlantic*, 23 September.
71 Department of Health and Social Care. 2021. Proposal for mandatory COVID
    certification in a Plan B scenario. Policy Paper, 27 September.
72 Nippon.com. 2020. The little-known story of the birth of the QR code. *Nippon.
    com*, 10 February. Scanova. Undated. What is a QR code? A beginner's guide.
    Blog Post, scanova.io.
73 Justin McCurry. 2020. Interview: 'I'm pleased it is being used for people's
    safety': QR code inventor relishes its role in tackling Covid. *The Guardian*,

11 December. European Patent Office. 2014. Masahiro Hara and team, winners of the Popular Prize. The little-known story of the birth of the QR code.

74   GOV.UK. Undated. Guidance. Create a coronavirus NHS QR code for your venue.

75   Beth Hale. 2020. Has Test and Trace become a stalker's charter? Giving your mobile number to a bar or restaurant is supposed to keep you safe but women are reporting troubling stories of texts from strangers and even harassment. *Daily Mail*, 7 October.

76   Rowland Manthorpe. 2021. Covid-19: Test and Trace barely used check-in data from pubs and restaurants – with thousands not warned of infection risk. *Sky News*, 4 March.

77   Ibid.

78   NHS Covid-19 app statistics: see 'Venue check-ins' and 'Venue alerts' tabs for week ending 21 July 2021.

79   BBC. 2021.Covid passports: how do they work around the world? *BBC News*, 26 July.

80   BBC. 2020. COVID-19: no plans for 'vaccine passport' – Michael Gove. *BBC News*, 1 December.

81   Oliver Milne. 2021. Matt Hancock warns tougher travel restrictions are on the way. *The Mirror*, 24 January.

82   BBC. Covid: minister rules out vaccine passports in UK. *BBC News*, 7 February.

83   Hannah Boland. 2021. Government funds eight vaccine passport schemes despite 'no plans' for rollout. *The Telegraph*, 24 January. Natalie Thomas. 2021. Vaccine passports: path back to normality or problem in the making? *Reuters*, 4 February. Stephanie Hare. 2021. Could Covid-19 vaccine passports use biometrics? *BBC Click*, 20 April.

84   Christopher Hope. 2020. Britons to get 'vaccine stamps' in their passports before overseas travel. *The Telegraph*, 28 November. Michael Cogley. 2021. Exclusive: vaccine passports to be trialled by thousands of Britons. *The Telegraph*, 12 January.

85   Chris Smyth and Henry Zeffman. 2021. NHS app to be converted for vaccine passports. *The Times*, 24 February.

86   Stephanie Hare. 2021. Give pause before you raise a glass to the prospect of a vaccine passport. *Guardian/Observer*, 28 March.

87   Sky. 2021. COVID-19: Boris Johnson rules out 'vaccine passports' for trips to pub. *Sky News*, 15 February.

88   BBC. 2021. Covid-19: pubs could require vaccine passports. *BBC News*, 24 March.

89   BBC. 2021. Covid: Jab passports possible 'after all offered vaccine'. *BBC News*, 25 March.

90   Reuters. 2021. Singapore to accept COVID-19 digital travel pass from next month. *Reuters*, 5 April.

91   Max Fisher. 2021. Vaccine passports, Covid's next political flash point. *New York Times*, 2 March.

92   Michael Gove. 2021. It's time to explore the need for Covid certification. *The Telegraph*, 3 April.

93   Ibid. See also: Telegraph Readers. 2021. Michael Gove asked Telegraph readers for their views on vaccine passports, here's what they had to say. *The Telegraph*, 5 April.

94 Graham Brady. 2021. Vaccine passports show the state is reaching too far into our lives. *The Telegraph*, 4 April. Maria Alvarez. 2021. Are Covid passports a threat to liberty? It depends on how you define freedom. *The Guardian*, 10 April.

95 Sarah Bosley. 2021. Covid-status certificates could lead to deliberate infections, scientists warn. *The Guardian*, 11 April.

96 Stefan Boscia and Poppy Wood. 2021. UK vaccine passport app could become a 'honeypot' for hackers, says former top government cyber adviser. *City AM*, 31 March.

97 Chris Smyth and Henry Zeffman. 2021. Hospitals and supermarkets to be exempt from Covid passport scheme. *The Times*, 1 April.

98 Aubrey Allegretti and Robert Booth. 2021. Covid-status certificate scheme could be unlawful discrimination, says EHRC. *The Guardian*, 14 April.

99 Ben Riley-Smith. 2021. Exclusive: we won't ask customers to show Covid passports, hospitality firms warn Boris Johnson. *The Telegraph*, 13 April.

100 James Robinson. 2021. GP surgeries, hospitals and supermarkets are set to be exempt from Covid vaccine passports as Boris Johnson prepares to announce more details of scheme on Monday. *Daily Mail*, 1 April.

101 Henry Zeffman. 2021. Covid contact app team NHSX now creating vaccine passports. *The Times*, 30 March.

102 BBC. 2021. Pub vaccine passports not British – Sir Keir Starmer. *BBC News*, 2 April.

103 Tony Diver. 2021. MPs to get vote on vaccine passports – and could defeat government in Commons. *The Telegraph*, 4 April.

104 Jessica Elgot and Aubrey Allegretti. 2021. UK Covid passports: how Boris Johnson's big plan fell flat. *The Guardian*, 18 June. House of Commons Public Administration and Constitutional Affairs Committee. 2021. No justification for Covid passports, say Committee. Report, 12 June. House of Commons Public Administration and Constitutional Affairs Committee. 2021. Covid-status certification. Second Report of Session, 10 June.

105 HM Government. 2021. COVID-status certification review. Report, July.

106 Christina Criddle. 2021. NHS app ready to become vaccine passport next week. *BBC News*, 11 May.

107 Harry Yorke and Ben Rumsby. 2021. Double jabs set to be needed to watch Premier League matches. *The Telegraph*, 24 July. George Bowden. 2021. Club night drops NHS Covid pass requirement. *BBC News*, 30 July.

108 Matthew Field. 2021. Covid passports risk Orwellian future of digital IDs, activists warn. *The Telegraph*, 17 July. Hayley Dixon, Lucy Fisher and Tony Diver. 2021. Vaccine passport firm says system could be 'redeployed' as a national ID card. *The Telegraph*, 13 July. Rob Davies. 2021. NHS app storing facial verification data via contract with firm linked to Tory donors. *The Guardian*, 15 September.

109 George Grylls and Kat Lay. 2021. Covid-19: Shapps backs businesses who want workers to get jabbed. *The Times*, 30 July.

110 NHSX website. Undated. App guidance for UEFA EURO 2020 ticket holders. Heather Stewart. 2021. Covid vaccine certificates to be compulsory for crowded venues in England. *The Guardian*, 19 July. Lucy Fisher. 2021. Boris Johnson urges use of Covid vaccine passports. *The Telegraph*, 12 July.

111   Aubrey Allegretti. 2021. Nudge or nutcracker? Either way PM faces vaccine passport backlash. *The Guardian*, 19 July.

112   Chris Smyth, George Ellis and Eleni Courea. 2021. Vaccine passports will coax people to get double jabbed, says Raab. *The Times*, 29 July.

113   Marie Jackson. 2021. England vaccine passport plans ditched, Sajid Javid says. *BBC News*, 12 September.

114   Chris Smyth. 2021. Sajid Javid won't rule out vaccine passports for pubs. *The Times*, 16 September.

115   BBC. 2021. Covid passports: how do I get one and when will I need it? *BBC News*, 1 October. BBC. 2021. Soft start for Scotland's vaccine passport scheme. *BBC News*, 2 October. BBC. 2021. Covid in Scotland: vaccine passport app launch hit by problems. *BBC News*, 1 October. Steven Morris. 2021. Wales to require NHS Covid passes to attend nightclubs and events. *The Guardian*, 17 September. Ruth Mosalski. 2021. Labour succeeds with plan to introduce Covid passes in Wales after farcical scenes in Senedd. *Wales Online*, 5 October. Claire Graham and Peter Coulter. 2021. Covid-19 NI vaccine cert app 'built for overseas travel only'. *BBC News*, 9 September. David Deans. 2021. Covid passes: Conservative who missed vote was at party conference. *BBC News*, 6 October.

116   BBC. 2021. Covid pass starts in England despite biggest rebellion of Johnson era. BBC News, 15 December. Matthew Field. 2021. How to get an NHS Covid Pass and what it allows you to do – from entering pubs to travelling abroad. *The Telegraph*, 15 December.

Conclusion

1   Taylor Downing. 2018. *1983: The World at the Brink*, pp. 195–200. London: Little Brown.

2   Casey Fiesler. 2018. Tech ethics curricula: a collection of syllabi. *Medium*, 5 July. Natasha Singer. 2018. Tech's ethical dark side: Harvard, Stanford and others want to address it. *New York Times*, 12 February.

3   J. Robert Oppenheimer's personnel hearings transcripts, volume XII. Cited in Ananyo Bhattacharya. 2021. *The Man from the Future: The Visionary Life of John von Neumann*, p. 212. London: Allen Lane.

4   John C. Camillus. 2008. Strategy as a wicked problem. *Harvard Business Review*, May. Horst Rittel and Melvin Webber. 1973. Dilemmas in a general theory of planning. *Policy Sciences* 4, 155.

5   Ibid., pp. 160, 163.

6   Apple. 2016. Q&A: answers to your questions about Apple and security: why is Apple objecting to the government's order. Jonathan Mayer and Anunay Kulshrestha. 2021. We built a system like Apple's to flag child sexual abuse material – and concluded the tech was dangerous. *Washington Post*, 19 August.

7   Paresh Dave and Jeffrey Dastin. 2021. Money, mimicry and mind control: Big Tech slams ethics brakes on AI. *Reuters*, 8 September.

8   Hetan Shah. 2021. Britain suffers from outdated thinking on innovation. *Financial Times*, 26 June.

9   Bruce Schneier. 2020. We're banning facial recognition. We're missing the point. *New York Times*, 20 January.

10 Michael S. Bernstein, Margaret Levi, David Magnus, Betsy Rajala, Debra Satz and Charla Waeiss. 2021. ESR: ethics and society review of artificial intelligence research. Preprint, arXiv (22 June).

11 Ian Sample. 2019. Maths and tech specialists need Hippocratic oath, says academic. *The Guardian*, 16 August.

12 Joseph Rotblat's interview is part of 'Voices of the Manhattan Project', US National Museum of Nuclear Science and History. Interview conducted on 16 October 1989. Sir Joseph Rotblat. 1999. A Hippocratic oath for scientists. *Science* **19**, 1475.

13 Karl Popper. 1968. Address delivered on 3 September to a special session titled 'Science and ethics: the moral responsibility of the scientist'. International Congress of Philosophy, held in Vienna. Further revised for publication in Karl Popper. 1994. *The Myth of the Framework: In Defence of Science and Rationality.* edited by M. A. Notturno. London/New York: Routledge.

14 Nonny Onyekweli. 2020. A medical school class thought the Hippocratic oath fell flat. So they wrote their own script. *Washington Post*, 26 September.

15 Tom Beauchamp and James Childress. 2001. *Principles of Biomedical Ethics*, 5th edn. Oxford University Press.

Glossary

1   Web Accessibility Initiative. Undated. Accessibility, usability, and inclusion. URL: www.w3.org/WAI/fundamentals/accessibility-usability-inclusion. See also Amélie Mourichon. 2021. What are the differences between universal design, accessibility, and inclusive design? *Say Yeah!*, 12 May.

2   World Bank. 2018. Belt and Road Initiative. Brief, 29 March. Lily Kuo and Niko Kommenda. 2018. What is China's Belt and Road Initiative? *The Guardian,* 30 July

3   Pasquale (2015). *The Black Box Society: The Secret Algorithms that Control Money and Information*, p. 3

4   Web Accessibility Initiative (undated). Accessibility, usability, and inclusion. See also Mourichon (2020). What are the differences between universal design, accessibility, and inclusive design?

5   Peter Foster, Madhumita Murgia and Javier Espinoza. 2021. UK suggests removing EU's human review of AI decisions. *Financial Times*, 9 September. Joe Mayes. 2021. UK may unleash AI bots online with no human safeguards. *Bloomberg*, 10 September. UK Department for Digital, Culture, Media & Sport. 2021. Data: a new direction. Guidance, 10 September.

Further reading

Below is a short list of suggestions for further reading. For a full bibliography of this book, please visit www.harebrain.co/books.

Baggini, Julian. 2018. *How the World Thinks: A Global History of Philosophy.* London: Granta.

Benjamin, Ruja. 2019. *Race after Technology.* Cambridge, MA: Polity.

Coeckelbergh, Mark. 2020. *Introduction to the Philosophy of Technology.* Oxford University Press.

Crawford, Kate. 2021. *Atlas of AI.* New Haven, CT/London: Yale University Press.

Criado Perez, Caroline. 2019. *Invisible Women: Exposing Data Bias in a World Designed for Men.* London: Chatto and Windus.

Farivar, Cyrus. 2018. *Habeas Data: Privacy vs. the Rise of Surveillance Tech.* New York/London: Melville House.

Friedman, Batya, and David G. Hendry. 2019. *Value Sensitive Design.* Cambridge, MA: MIT Press.

Juma, Calestous. 2016. *Innovation and Its Enemies: Why People Resist New Technologies.* Oxford University Press.

Mullaney, Thomas S., Benjamin Peters, Mar Hicks and Kavita Philip. 2021. *Your Computer Is On Fire.* Cambridge, MA: MIT Press.

O'Neil, Cathy. 2016. *Weapons of Math Destruction.* London: Allen Lane.

Perlroth, Nicole. 2021. *This Is How They Tell Me the World Ends: The Cyber Arms Race.* London. Bloomsbury.

Russell, Stuart. 2019. *Human Compatible: AI and the Problem of Control*. London: Allen Lane.

Schneier, Bruce. 2015. *Data and Goliath: The Hidden Battles to Collect Your Data and Control Your World*. New York: W. W. Norton.

West, Darrell M., and John R. Allen. 2020. *Policymaking in the Era of Artificial Intelligence*. Washington, DC: Brookings.

Acknowledgements

It is a relief to thank publicly the many people who helped me to publish this book. Before I do so, though, it goes without saying that responsibility for any errors herein is mine alone. The publishers and I will be pleased to correct any errors in future editions.

Professor Diane Coyle invited me to write this book for her Perspectives series, and Richard Baggaley and Sam Clark at London Publishing Partnership helped me to get to the finished article that is in your hands now. I am grateful for their confidence, patience and editing.

Noma Bar created the art that graces this book's covers, and Joel Minter and Helen Cowley at Dutch Uncle helped make it happen. Julia and Maria-Luisa Puig invited me to the event where we first met Noma – just one of the many ways in which they (and Benjamin Huaman de los Heros) make life better.

Jeremy Smith worked with me to create the original graphics and data visualizations in this book, and his skill and humour enriched the project. We both thank Alex Selby-Boothroyd for a helpful discussion about the official UK data on Covid-19 hospital admissions. Dr Jennifer Fraser introduced me to Kyra Araneta, who assisted with the bibliography; to Terese Jonsson, who helped with formatting; and to Aislin Baker, who gave really helpful feedback on the first full draft. Dr Shehnaz Suterwalla invited me to present some early thinking to her students at the Royal College of Art, and their questions and comments pushed me to consider new ideas.

ACKNOWLEDGEMENTS

Randall Munroe of XKCD.com allowed me to use two of his comics in the book ('Free Speech' and 'CAPTCHA') and Casey Blair helped to make that happen. Dr Dawn Alexandrea Berry allowed me to use her photos of the panopticon Presidio Modelo in Cuba. Shanti Das and the *Sunday Times* gave permission for me to share their reporting and their graphic on the partnership between the police and Amazon Ring in the United Kingdom.

The London Library posted me books during lockdowns, while the staff at Daunt Books on South End Green and Haverstock Hill kept me well stocked with the latest titles.

For discussions on facial recognition technology and the history of cybersecurity I thank Professor Paul Wiles, Norman Lamb, Lord Clement-Jones, Lord Scriven, Silkie Carlo, Pippa King, Jen Persson, Liza Mundy, Dr Mar Hicks, Jason Fagone, Giles Herdale and Detective Superintendent Bernie Galopin, and I thank Azeem Azhar and Marija Gavrilova for the opportunity to guest edit the *Exponential View* newsletter on biometric and surveillance technologies. My deep thanks also to Tabitha Goldstaub and Charlie Muirhead of CognitionX.

For discussions on the pandemic and the digital health tools it inspired, I thank Dr Aminah Sadiah Verity, Dr Naeem Ahmed, Dr Winta Mehtsun, Dr Richard McKay, Dr Alexa Hagerty, Dr Yewande Okuleye, Dr Alexandra Albert, Dr Andrew Trathen and his colleagues in Hackney public health, Professor Susan Michie, Professor Danny Altman, Simon Binns and Andrew Bud.

Many people took time to critique this book at various stages while working their own full-time jobs; caring for children, elders and pets; moving house; going to A&E for various ailments and injuries; falling ill with Covid-19 and otherwise coping with the pandemic, Brexit, the January 6 insurrection, the collapse of the Texas electricity grid ... and more. David Matthew Millar, Ankur Banerjee, Lyndsay Baker and James Dickinson read early thinking and later thinking, talked through ideas at all hours, and challenged me where necessary – I cannot thank them enough. Dr Danny Moore, Professor Peter Fussey and Ely Cossery read

selected chapters and improved them. The first full draft ben-
efitted from the critique of Richard Allen Greene, Dr Richard
McKay, Dr Alexa Hagerty, Dr Chris Oates, Nik Speller, Michael
Bruce, Lord Clement-Jones, Catherine Miller, Dr Paul Killworth,
Ann Finkbeiner and Dr Robert Boyce.

I am grateful for the education, training and opportunities
that made it possible to write this book. Thank you to my teachers in St. Charles, Illinois, and to my professors at the University
of Illinois at Urbana-Champaign, the Université de la Sorbonne
(Paris IV) and the London School of Economics. The Arts and
Humanities Research Council funded much of my PhD research,
with support from the Society for the Study of French History,
the LSE International History department, and the University
of London Central Research Fund. At the University of Oxford,
St Antony's College awarded me a visiting fellowship for a year
of pure research, and St Catherine's College, St Hugh's College
and Jesus College gave me the chance to hone my teaching
skills. I also learned a great deal from my bosses, colleagues
and clients at Accenture, Oxford Analytica, Palantir and Accenture Research, as well as those with whom I have worked since
becoming an independent researcher. Thank you.

In 2013 the BBC selected me for its Expert Women programme, which during the years since has given me the chance
to learn from Rory Cellan-Jones, Jat Gill, Leo Kelion, Zoe Kleinman, Jane Wakefield, Osman Iqbal, Omar Mehtab, Aaron Hes-
elhurst, Sally Bundock, Susannah Streeter, Lucy Burton, Nishi
Banga, Ania Lichtarowicz, Gareth Mitchell, Ghislaine Bodding-
ton, Ben Brown and Oliver Woods.

Thank you to Elmira Bayrasli and Lauren Bohn for awarding
me a Foreign Policy Interrupted fellowship, which set me on a
path to writing op-eds, and thank you to the editors who published those op-eds: Jonathan Derbyshire at the *Financial Times*,
Robert Yates at *The Observer*, Christian Caryll at the *Washington
Post*, David Baker at *WIRED*, and Kate Bevan at *Which? Computing*. Thanks, too, to Anne-Marie Slaughter, my co-author in an

op-ed for *Project Syndicate*. I learned a lot from the journalists Madhumita Murgia at the *Financial Times*, Ali Rae at *Al-Jazeera*, Parmy Olson at the *Wall Street Journal* and now Bloomberg, and Rob Davies at *The Guardian*.

Daniel Rix and Tony Veitch of Specialist Speakers Agency, Ben Cherwell and his team at Davis Grant, and Gary Pomeroy and Jana Helmuth helped to keep me in business. Chrissie Mullings-Lewis, Naomi Annand and Katrina Kurdy gave me another perspective on the 'mind–body problem' with weights and yoga. Thank you for helping me to stay strong throughout the writing of this book. Noj and his team at Little Oak Coffee kept our neighbourhood caffeinated and bantering even in the darkest days of the pandemic.

For their friendship I thank the aforementioned Maria-Luisa and Benja (Julia), Jeremy, Jennifer, Aislin, Shehnaz and Azeem (Salman, Sophie and Jasmine), Dawn, Pippa, Rich, Alexa, David (Ellie), Ankur and Ely (Daisy), Lyndsay, James, Danny, Richard, Chris, Nik, Mike and Dr Robert Boyce. I also thank Hugo Warner and Maria João Paixão (Teresa and Beatriz), Katie Dunn (Truffle), William Atwell and Matt Lecznar, Mary Greer, Jim and Sophie Copeman, Genevieve Cuming, Sophie and John Power (Donnacha, Cormac and Saoirse), Eduardo Plastino, James Barr, Graham Ball, Dr Mara Tchalakov, Benjamin Charlton, Andy Him, Jason Crabtree, Lisa Dittmar, Dr Maria Chen, Dr Magdalena Delgado, Professor Rana Mitter and Professor Margaret MacMillan.

Finally, thank you to my family: Gene, Amanda, Jordan, Jackson and Jason; Amber and Tom; Jan and Rick; and especially my parents, Sharon and Eugene – first readers, first teachers, first love.

About the author

Stephanie Hare is an independent researcher and broad-caster focused on technology, politics and history. Selected for the BBC Expert Women programme and the Foreign Policy Interrupted fellowship, she contributes frequently to radio and television and has published in the *Financial Times*, the *Washington Post*, *The Guardian*, *The Observer*, the *Harvard Business Review*, and *WIRED*. Previously she worked as a principal director at Accenture Research, a strategist at Palantir, a senior analyst for Western Europe at Oxford Analytica, the Alistair Horne Visiting Fellow at St Antony's College, Oxford, and a consultant at Accenture.

She holds a PhD and an MSc from the London School of Economics and a BA in Liberal Arts and Sciences from the University of Illinois at Urbana-Champaign, including a year at the Université de la Sorbonne (Paris IV).

Figures

Index

Bluetooth, 158, 173–4
bone, 24
border technology, 117–18
Boris Johnson, 140
Boulogne, Guillaume-Benjamin-Amand Duchenne de, 95
Bowie, David, 32–4
Boyle, Katherine, 79
brain implants, 47
brand loyalty, 71
Brexit, 58
Bridgers, Sarah, 51
Bridges, Ed, 138
British Computing Society, 5
British Security Industry Association, 141
British Transport Police, 143
broad release, 77–8
Broussard, Meredith, 43, 45
Browne, Simone, 108
Bryson, Joanna, 39
Budgens, 141
Buolamwini, Joy, 125–6
Bush, Vannevar, 25
Business Research Company, 91

C

calligraphy, 73
Cambridge University, 5
cameras; automatic number plate recognition, 141; body worn video, 126, 141; CCTV, 117, 141, 143; infrared, 117; unmanned aerial vehicles (drones), 117, 141; vehicle borne, 141
Camillus, John C., 199
Canada, 81, 97
Capitol, 1, 5
CAPTCHA, 65–7
Carlo, Silkie, 132, 134
Carnegie Endowment for International Peace, 91, 114
Carnegie Mellon University, 196
Cartesian dualism, 41

CCTV, 117, 141, 143
Cellan-Jones, Rory, 174
censorship, 60, 79
Center for Privacy and Technology at Georgetown Law, 128
Centers for Disease Control and Prevention, 169
Central Intelligence Agency (US), 48
Centres for Doctoral Training, 196
chain of transmission, 160, 173, 181, 189
Charlottesville white nationalist rally, 68
Chauvin, Derek, 121, 126
chemistry, 7
child sexual abuse, 17
Chile, 157
China, 16–17, 70, 113; Belt and Road Initiative, 80, 114; Chinese Communist Party, 68; gig workers, protection for, 5; Project Dragonfly (Google), 59; rivalry with US, 81; settings of international technology standards, 80; Uyghur muslims, persecution of, 68, 81, 121, 146, 152–5
Chomborazo (volcano), 29
Churchill, Winston, 87
citizen science, 169
City of London Corporation, 142
civil liberties, 9, 16, 31, 69–70, 89, 128–9, 135, 138, 140, 151–2, 157, 182, 202
Clarke, Yvette, 108
Clearview AI, 96–8, 112
Clement-Jones, Lord Timothy, 106, 140
climate change, 197
climate zones, 30
closed-circuit television cameras. See CCTV
Cloudflare, 68
CloudWalk, 114